T0275962

SpringerBriefs in Microbiology

More information about this series at http://www.springer.com/series/8911

Deepansh Sharma · Baljeet Singh Saharan
Shailly Kapil

Biosurfactants of Lactic Acid Bacteria

 Springer

Deepansh Sharma
School of Biosciences
Lovely Professional University
Phagwara, Punjab
India

Baljeet Singh Saharan
Department of Microbiology
Kurukshetra University
Kurukshetra, Haryana
India

Shailly Kapil
Division of Dairy Microbiology
National Dairy Research Institute
Karnal, Haryana
India

ISSN 2191-5385   ISSN 2191-5393   (electronic)
SpringerBriefs in Microbiology
ISBN 978-3-319-26213-0   ISBN 978-3-319-26215-4   (eBook)
DOI 10.1007/978-3-319-26215-4

Library of Congress Control Number: 2016932323

Printed on acid-free paper

This Springer imprint is published by SpringerNature
The registered company is Springer International Publishing AG Switzerland

In Loving Memories of my Mother

Late. Smt. Kunta Sharma

Foreword

Biosurfactants, microbial surfactants, and emulsifiers have attracted considerable attention in the past years and they have been investigated extensively as an alternative for chemically synthesized conventional surfactants. Typically, their hydrophobic moiety is composed of saturated or unsaturated fatty acids, hydroxy fatty acids, or fatty alcohols, whereas the hydrophilic moiety often contains a mono-, oligo-, or polysaccharide, an amino acid, peptide, or protein. The utilizable physicochemical properties of biosurfactants are generally comparable to chemically synthesized compounds. Advantageously, biosurfactants can be produced from renewable resources or industrial waste products and are biodegradable. In addition, several biosurfactants have been reported to have manifold biological activities.

Many of our most consumed foods, toiletries, cosmetic products, and pharmaceuticals are emulsions. Therefore, the range of applications for surfactants and emulsifiers is huge. However, the application of alternative biosurfactants in innovative products has only emerged in the past decade. Today only rhamnolipids, sophorolipids, and mannosylerythritol lipids are among the few commercially relevant biosurfactants. Although they have been known for several decades and their production is technically feasible, they are confined to small business niches and no broad applications are foreseeable in the near future. Nevertheless, there is a demand for new and innovative surfactants. Especially in the food industry, new and safe surfactants are desired as there is only a limited number of surfactants and emulsifiers approved by now.

What may be the reason for this situation? First, microbial production seems to be not competitive by economical aspects and the development of more competitive processes is a challenging task. A second reason may be the lack of availability of commercial samples for the development of novel applications. This may change, however, in view of the many reported advantages of biosurfactants.

In my opinion, the one most disremembered advantage of biosurfactants is the wide variety of chemical structures and their broad range of physical properties covered. The scientific development in this area has witnessed a renaissance, as it becomes evident that the number of microorganisms producing surfactants and also the molecular diversity of microbial surface active substances may both be more diverse than previously expected. The book now presented by Deepansh Sharma, Baljeet Singh Saharan, and Shailly Kapil illuminates this for the biosurfactants derived from lactic acid bacteria, a little investigated group in this regard. The book represents the first focused monograph on this issue presenting valuable information as functional properties and structural composition.

The read shows that research opportunities are still ubiquitous in this field. The identification and characterization of new biosurfactants, the isolation of new lactic acid bacteria, the biosynthesis, and its genetic organization are just the most obvious research questions. Bioprocessing, purification, and application development being subsequent topics.

One expectation of biosurfactants derived from overproducing and non-pathogenic wild-type lactic acid bacteria is that these will qualify as alternatives for food surfactants eventually.

I wish all of you an enjoyable read full of inspiration.

Prof. Dr. Rudolf Hausmann
Head
Faculty of Natural Sciences Institute of Food Science and
Biotechnology Department Bioprocess Engineering

Preface

The idea of writing a book on "**Biosurfactants of Lactic Acid Bacteria**" struck us instantaneously after I was enrolled as a Ph.D. scholar in Microbiology and start working on biosurfactants derived from lactic acid bacteria. After five years of research related to the biosurfactants obtained from lactic acid bacteria, I and my mentor decided to compile all the information as a small compendium on "Biosurfactants of Lactic Acid Bacteria." And we found Springerbrief series as the most appropriate way to publish our compendium. The prompt and positive response from the Springer team through their valuable suggestions and timely contributions are gratefully appreciated. The book consists of 7 chapters on different aspects; each one represents the progress, prospect, and challenges in biosurfactant of lactic acid bacteria research. This is supposed to be the most up-to-date book on "Biosurfactants of Lactic Acid Bacteria." We attempt not only to highlight the remarkable progress made by the scientific community in this field of research, but also to critically examine the lacuna to expand the commercial prospects of these wonder biomolecules.

The term biosurfactant refers to those compounds that have surface tension active properties, the molecules that reduce interfacial tension. The chemical composition of biosurfactants can vary widely, but they have in common their amphiphilic or amphipathic nature. These features make biosurfactants advantageous in a wide variety of industrial formulations based on their capabilities to lower surface tensions, increase solubility, their detergency power, wetting ability, and foaming capacity. First, biosurfactants are considered environmentally "friendly" since they are moderately nontoxic and biodegradable. Second, biosurfactants have exceptional structures that are just starting to be cherished for their potential applications to industrial biotechnology to environmental cleanup. This "Biosurfactants of Lactic Acid Bacteria" book covers the current knowledge and the most recent advances in the field of microbial biosurfactants. The book includes the physicochemical properties of biosurfactants, their role in the physiology of the microbe that produced them, the biosynthetic pathway for their production, including the genetic regulation, and their potential biotechnological applications.

I thankfully acknowledge all the co-authors of each chapter of the book for their valuable and inspiring contributions, especially my Mentor "Dr. Baljeet Singh Saharan" for giving me this opportunity. I do highly appreciate the help that I have constantly received from my colleagues at Springer Germany and India. I also thank my loving wife, Deepti Singh, for her constant support and patience. I also thank my research students, particularly Gurkiran Parmar, Sakshi Sood, Harsimran Kaur, Sonali, and Sandeep Singh, for their technical support, understanding, and forbearance.

Deepansh Sharma

Contents

1 Introduction.. 1
    Lactic Acid Bacteria-from Gut to Fermentation 1
    Biosurfactants (Surface Active Agent) 2
    Why the Microbial Cells Produce Biosurfactant, Significance,
    and Their Role to the Cell?................................. 3
    Classification of Biosurfactants 4
    Application of Biosurfactants 4
    Antimicrobial Activity 6
    Anti-biofilm Nature 7
    Future Prospects 9
    References 10

2 Biosurfactants of Probiotic Lactic Acid Bacteria 17
    Introduction 17
    LAB-Derived Biosurfactants in Biomedical.................... 21
    LAB-Derived Biosurfactants as Antimicrobials 23
    Conculusion 25
    References 26

3 Properties of Biosurfactants 31
    Introduction 31
    Surface Tension... 32
    Critical Micelle Concentration............................... 33
    Emulsification... 37
    Stability 40
    Toxicity 42
    Phytotoxicity.. 44
    Conclusion 44
    References 45

4   Structural Properties of Biosurfactants of Lab   47
     Introduction   47
     Glycolipids   49
     Rhamnolipids   49
     Trehalolipids   49
     Sophorolipids   50
     Lipopeptides and Lipoproteins   50
     Fatty Acids and Phospholipids   50
     Polymeric Biosurfactants   51
     Structural Properties of Biosurfactant of LAB   51
     NMR and FTIR Analysis of LAB Derived Biosurfactants   53
     Fatty Acids   56
     Conclusion   57
     References   58

5   Substrates and Production of Biosurfactants   61
     Introduction   61
     Agro-Industrial Substrates   64
     Biosurfactant Production from Cheese Whey   64
     Biosurfactant Production from Lignocellulosic Waste   66
     Biosurfactant from Synthetic Medium   67
     Other Miscellaneous Substrates   68
     Conclusion   68
     References   69

6   Applications of Biosurfactants   73
     Introduction   73
     Antimicrobial Potential of Biosurfactants   74
     Antibacterial Activity   75
     Anti-adhesion Activity   78
     Conclusion   79
     References   79

7   Future Prospect   83
     Introduction   83
     Biosurfactant Mediated Synthesis of Nanoparticles   84
     Biosurfactants in Animal Feed and Food   85
     Biosurfactants in Cosmetics   85
     Conclusion   85

Chapter 1
Introduction

Abstract The Lactic acid bacteria (LAB), commonly associated with food and feed fermentation normally be inherent in the mucosal surfaces of healthy humans and animals. Microbial surface active agents are amphiphilic compounds produced commonly by microorganisms predominately bacteria and yeast on their cell surface, or extracellularly with exceptional surface and emulsifying activities. The physiological function of microbial surfactants in a producer cell is not entirely understood. On the contrary, there has been hypothesis about their involvement in emulsification of water insoluble substrates. Dissimilar to chemical surfactants, which are categorized according to the nature of their polar grouping, biosurfactants are categorized largely by their chemical composition and their microbial origin. Recent advances in biological disciplines and analytical approaches have focused about the enormous rise in biosurfactant for applications in environmental, bio medicine, food/feed, and cosmetics industries. The demand for novel biosurfactants in the cosmetics, food and pharmaceutical formulations, is progressively increasing and the biosurfactants with effective and eco-friendly composition, impeccably meet this demand. Most of the biosurfactant-producing microorganisms are pathogenic and challenging to handle in commercial formulations. The development of biosurfactant production from nonpathogenic microorganisms such as "Biosurfactant derived from LAB" is a prevailing task that is receiving increased attention in direction to escape pathogenicity. Detailed studies of their natural roles in microbial interactions, cell signaling, pathogenicity, and biofilm development could advocate significant future applications.

Keywords Lactic acid bacteria · Biosurfactants · Antibiofilm · Antimicrobial and glycolipids

Lactic Acid Bacteria-from Gut to Fermentation

Lactic acid bacteria (LAB) comprise gram-positive bacteria grouped according to their morphological, physiological, and metabolic features. The LAB commonly associated with food and feed fermentation, is normally associated with the mucosal

© The Author(s) 2016
D. Sharma et al., *Biosurfactants of Lactic Acid Bacteria*,
SpringerBriefs in Microbiology, DOI 10.1007/978-3-319-26215-4_1

surfaces of healthy humans and animals (Ouwehand et al. 2002; El Aidy et al. 2015). LAB, which are generally regarded as safe (GRAS, the United States Food and Drug Administration), are recognized for application in the food and health sector. Attention in the valuable effects of LAB dates to the Russian microbiologist, Elie Metchnikoff, who demonstrated that the extended longevity of the Balkan people could be due to their habit of ingesting fermented milk products. He assumed that gastrointestinal tract inhabitant by LAB prevents the harmful effects of putrefying microbes.

LAB comprises a wide range of genera and includes a considerable number of species of different lactic acid bacteria. Bacterial species belonging to the genus *Lactobacillus* are part of the human and animal commensal intestinal flora predominantly (Douillard 2014). In the past decades, there has been an increasing recognition of the role of LAB in the maintenance of the homeostasis within the ecosystem of gastrointestinal tracts and in combating colonization of pathogens (Ghosh et al. 2014). Some lactobacilli play protective roles by producing compounds such as lactic acid, bacteriocins, hydrogenperoxides, and biosurfactants, which inhibit the growth of potential pathogens (Agboola et al. 2014).

Biosurfactants (Surface Active Agent)

Microbial surface active agents are amphiphilic compounds produced commonly by microorganisms predominately bacteria and yeast on their cell surface, or extracellularly with exceptional surface and emulsifying activities (Maneerat 2005). Biosurfactant contains both hydrophobic (e.g., long chain fatty acids, hydroxy fatty acids) and hydrophilic (e.g., carbohydrate, amino acids, peptides, carboxylic acid, alcohol, phosphate) moieties produced which accumulate at the interface between liquid phases, and therefore can reduce the surface/interfacial tension in liquids (Saharan et al. 2011; Banat et al. 2010; Van Hamme et al. 2006). In the past decades biosurfactants have attracted attention due to their less toxic, diverse and biodegradable, highly selective, specialized functions unlike synthetic surfactants (Banat et al. 2010; Nitschke et al. 2005).

Microbial surfactants also have been found to be very significant at low amount and stable to environmental circumstances such as wide range of pH, extreme temperature ranges, salinities, alkaline and acidic environments. Biosurfactants showed superior environmental compatibility, lower critical micelle concentration, higher specificity, and can be synthesized from renewable low-cost resources especially agoroindustrial wastes (Desai and Banat 1997; Oliveira et al. 2009; Rahman and Gakpe 2008). The global surfactant production has surpassed 2.5 million tons in 2002 and was nearly 1735.5 million USD in the year 2011. Surfactant production is estimated to reach 2210.5 million in the year 2018, with an average yearly growth rate of 3.5 % from 2011 to 2018 (Sekhon et al. 2012;

Van Bogaert et al. 2007). Additionally, commonly used synthetic surfactants are unsafe, poorly biodegradable and may lead to the buildup of ecologically detrimental compounds in soil and water bodies (Kuyukina et al. 2005; Desai and Banat 1997; Batista et al. 2006. In addition, the production of synthetic surfactant contributes to the exhaustion of the global petrochemical resources.

Why the Microbial Cells Produce Biosurfactant, Significance, and Their Role to the Cell?

The physiological function of microbial surfactants in a producer cell is not entirely understood. On the contrary, there has been hypothesis about their involvement in the emulsification of water-insoluble substrates. When microbial cells are cultivated, oils and growth-stimulating substances are frequently accumulated in the production medium. These accumulated substances have roles in emulsifying the substrate, spreading the interfacial area and expediting mass transfer on the surface of cell. The interfacial surface between aqueous phase and oil can be a restrictive factor and emulsification is a natural process advantageous for the microbial cell to absorb the substrate (Toren et al. 2002; Hua et al. 2003).

In other cases, when pathogens infect plants or animals, biosurfactants are considered to function as a dispersing and wetting agent for the surface of the host cells (Matsuyama et al. ; Hildebrand et al. 1998; Matsuyama and Nakagawa 1996). On the other hand, biosurfactants play a significant role in adhesion of the cells to the interfaces. Adhesion is a physiological mechanism for survival of microbial cells in the environment (Stoodley et al. 2004; Dunne 2002). Microbial cells can also use their biosurfactants to regulate their cell surface properties like to attach or detach from biotic and abiotic surfaces (Rosenberg and Ron 1997, 1999).

Physiologically biosurfactants have been recognized as the components of cellular metabolism, motion, and defense. They have been reported majorly in bacteria, biofilms, as a molecule for quorum-sensing, encouraging the uptake of poorly soluble substrates, lubricants, immune modulators, secondary metabolites, and known antimicrobial substances (Fracchia et al. 2012; Dusane et al. 2010).

Microbial surfactants act as vital molecules for interfacial processes, which conditioned the cell surface, with which the cells interact (Neu 1996). Biosurfactants also have significant roles in the dissolution of oil molecules particularly for oil-degrading microorganisms (Ron and Rosenberg 2002). The role of bacterial biosurfactants has been widely studied in *Pseudomonas* where they are known to stimulate colonization (Pamp and Tolker-Nielsen 2007).

Fig. 1.1 Classification of
biosurfactants

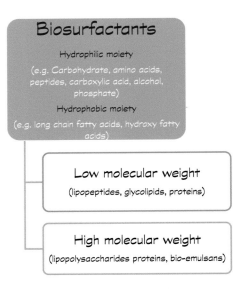

Classification of Biosurfactants

Dissimilar to chemical surfactants, which are categorized bestowing to the behavior
of their polar grouping, microbial surfactants are classified majorly by their
chemical constituents and the microorganism produced (Fig. 1.1). Commonly,
biosurfactants structure is composed of a hydrophilic moiety consisting of amino
acids, peptides, anions, or cations; polysaccharides; and a hydrophobic moiety
composed of unsaturated, saturated, or fatty acids (Saharan et al. 2011).

In view of that, the major group of biosurfactants comprises phospholipids, fatty
acids, glycolipids, lipopeptides, lipoproteins, polymeric surfactants, and particulate
surfactants.

Application of Biosurfactants

Recent advances in biological disciplines and analytical approaches have focused
about the enormous rise in biosurfactant for applications in environmental, bio-
medicine, food/feed, and cosmetics industries (Banat et al. 2010; Perfumo et al.
2010). Environmental contamination due to industrial leftover is due to the unin-
tentional or deliberate discharge of organic or inorganic substances into the natural
environment. Such components pose complications for remediation, as they
become attached to soil particles. The application of biosurfactants aims at growing
their availability or mobilizing contaminants.

Several biosurfactant exhibits antibacterial, antifungal, strong surface activity, emulsion forming ability, antioxidant, moisturizers, antiradical, antibiofilm, anti-tumor, oil-recovery, nanotechnology, stabilizing agents, and antiviral activities, making them appropriate candidates to combat infections (Singh and Cameotra 2004; Nitschke et al. 2010; Ongena and Jacques 2008; Rosenberg and Ron 1999; Banat et al. 2000; Rahman et al. 2002). Microbial surfactants have been explored for a variety of industrial and environmental applications, pharmaceutical and food processing industries (Fig. 1.2) (Bodour et al. 2003; Desai and Banat 1997; Makkar and Cameotra 1998; Banat et al. 2010; Saharan et al. 2011; Kiran et al. 2010; Khopade et al. 2012; Kryachko et al. 2013; Ławniczak et al. 2013; Sachdev and Cameotra 2013).

In aqueous phase, the interfacial action of microbial surface active agents can allow substrates like hydrocarbons more responsive to the degradative capabilities of the microbial cell (Kostka et al. 2011). The mechanisms which elaborated the interactions stuck between biosurfactants and the microbial cells include the following (Fuentes et al. 2014):

- Emulsification
- Adhesion/de-adhesion of microorganisms

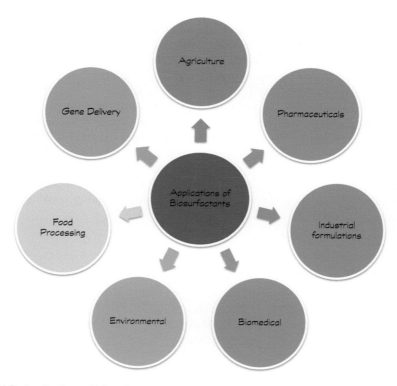

Fig. 1.2 Applications of biosurfactants

- Micellarisation
- Desorption of contaminants.

In demand to enhance the degree of hydrocarbon degradation, inorganic nutrients, and microbial surfactants can be augmented to the system (Ferradji et al. 2014). Maximal degradation magnitude of hydrocarbons was attained when the inorganic nutrient additions and biosurfactant were augmented cooperatively (Rahman et al. 2003; Banat et al. 2010). Biosurfactants also have widespread application in the petroleum sector. Microbially enhanced oil recovery (MEOR) is a practice that whichever uses a crude biosurfactant preparation or a whole killed microbial cell culture to release crude oil from a binding substrate (Perfumo et al. 2010; Banat et al. 2010).

A practice of MEOR has been established efficiently to recover oil from the sludge that accumulates in oil storage tanks (Banat et al. 1991). The application of biosurfactants to eliminate contaminants from soils is a method developed, i.e., biosurfactant-mediated bioremediation, since the exclusion efficacy is typically driven by the chemico-physical properties of the microbial surfactants. The mechanisms affecting hydrocarbon mobilization be similar to those convoluted in improving bioavailability for bioremediation (Perfumo et al. 2010; Banat et al. 2010).

Biosurfactants can stimulate the growth of producing microbial cells on hydrocarbons by accumulating the surface area between oil and water or emulsification (Miller and Zhang 1997). High-molecular-surface active agents (bio-emulsifiers) have enormous prospective for calming emulsions flanked by hydrocarbons and water, thus increasing bacterial biodegradation. In case of low-molecular-weight biosurfactants, intensify the extent of bioavailability of contaminants by degrading microorganisms (Mulligan 2009).

Due to their various useful physicochemical characteristics, the use of biosurfactants has also been anticipated for several industrial applications, as food additives, cosmetics formulations, and detergents preparations (Banat et al. 2010, 2000). In the food processing industry, the most advantageous property, it improves the texture and creaminess of dairy products. Surface active agents are also used to hinder staling, solubilize flavor oils, and to develop organoleptic feel in ice cream formulations, bakery preparations, and as fat stabilizers during cooking of fats (Nitschke et al. 2007).

Antimicrobial Activity

The antimicrobial potential of various microbial surfactants has been documented in the literature for diverse applications (Banat et al. 2010; Cameotra and Makkar 2004). A number of biosurfactants are recognized to encompass various therapeutic applications (Khire and Khan 1994; Banat 1995; Peypoux et al. 1999). The exploration for novel antimicrobial drugs remains a foremost apprehension at the present time because of the recently emerged pathogens which become unresponsive to conventional antibiotics. Biosurfactants have been found to be appropriate alternatives to traditional antimicrobial compounds and may consequently be used as effectual and safe therapeutic molecules (Saharan et al. 2011; Banat et al. 2010, 2000; Singh and Cameotra 2004).

Biosurfactant can infiltrate into the plasma membrane through hydrophobic interfaces, thus impelling the hydrocarbon chains (Carrillo et al. 2003; Deleu et al. 2008; Bonmatin et al. 2003). Plasma membrane disruptions by biosurfactants are generic and are beneficial for accomplishment on the plasma membranes of bacteria (Lu et al. 2007; Rodrigues and Teixeira 2010; Das et al. 2008).

The antimicrobial biosurfactants include iturin and surfactin derived from *Bacillus subtilis* (Walia and Cameotra 2015; Alvarez et al. 2012; Ahimou et al. 2000), rhamnolipids from *Pseudomonas aeruginosa* (Benincasa et al. 2004) mannosylerythritol lipids from *Candida antarctica* (Morita et al. 2013; Arutchelvi et al. 2008), from probiotic bacteria *Lactobacillus casei* MRTL3 (Sharma and Saharan 2014) *Streptococcus thermophilus* A, and *Lactococcus lactis* (Rodrigues et al. 2004, 2006).

Antiviral potential of biosurfactants have also been documented (Naruse et al. 1990). Antiviral potential of biosurfactants consequently may take situate as viral lipid envelope and capsid breakdown leads to ion channels creation (Seydlová and Svobodová 2008). On the other hand, sophorolipids and trehalose lipids are also reported for anitiviral activity against human immunodeficiency virus and inducing proliferation of T-lymphocytes (Shah et al. 2007). Mycoplasma infectivity in cell culture experiments is a recurrent problem. Surfactin addition to the mammalian cell culture permitted definite inactivation of mycoplasmas devoid of negative effects on cell metabolism in the culture (Kumar et al. 2007; Vollenbroich et al. 1997).

The antifungal properties of microbial surfactants have been recognized against various food and super facial funal pathogens (Abalos et al. 2001). The biosurfactants produced by *Pseudozyma flocculosa*, was reported for in vitro antifungal potential against various pathogenic yeasts, together with *Cryptococcus neoformans*, *Trichosporon asahii* and *Candida albicans* (Mimee et al. 2005). Antifungal potential of biosurfactants against phytopathogenic fungi has similarly been reported. Cellobiose lipids, rhamnolipids, surfactin, iturin have been reported for varying degrees of antimicrobial activities (Kulakovskaya et al. 2009; 2010; Debode et al. 2007; Banat et al. 2010; Tran et al. 2007, 2008; Chen et al. 2009; Grover et al. 2009; Mohammadipour et al. 2009; Snook et al. 2009).

Anti-biofilm Nature

Microbial biofilms establishment on biomedical equipment (urinary catheters, central venous catheters, heart valves, voice prostheses, contact lenses) is a vital and risky incidence, chiefly as the bacteria in such microbial biofilms turn out to be extremely resistant to conventional antibiotics (Dettenkofer et al. 2005; Falagas and Makris 2009; Donlan and Costerton 2002; Stickler 2008; Petrelli et al. 2006; Litzler et al. 2007; Buijssen et al. 2007; Imamura et al. 2008). Microorganisms within biofilms evade host defenses and endure antimicrobial chemotherapy (Morikawa 2006; Falagas and Makris 2009).

Microorganisms in general incline near to surfaces developing biofilms as an approach to shield themselves from environmental stresses. Microbial cells are capable of sensing their specific cell density, converse, and perform cell-to-cell signaling such as quorum-sensing (Davies 2003). Microbial biofilms composed of one species are reasonably rare as compared to the composite multispecies populations (Stoodley et al. 2002; Bueno 2014; Kotulova and Slobodnikova 2010).

Microbial biofilms are the clumps of microbes attached and growing on a surface (Parsek and Singh 2003). The biofilm development is a stepwise progression mainly involving initial attachment, early progress of architecture, development, and dispersion (Fig. 1.3). The microorganism's preliminary attachment can be active or passive, mainly reliant on their motility and gravitational conveyance of free floating cells (planktonic) (Kumar and Anand 1998). Initially, the attached cells secrets only a small amount of extracellular polymeric substance (EPS) which supports to reinforce the bond between the bacteria and the substratum (O'Toole and Kolter 1998; Chmielewski and Frank 2003; Donlan and Costerton 2002). The adhesion of microbial cell to the surface is also reliant on the physicochemical characters of the surface such as texture, i.e., charge, hydrophobicity, pH, temperature, and nutrient content of the preconditioning solution (Nilsson et al. 2011; Abdallah et al. 2001). Microorganisms in natural systems have revealed that accumulation could clinch the enlistment of planktonic cells from the nearby medium as a consequence of cell-to-cell communication, i.e., quorum sensing

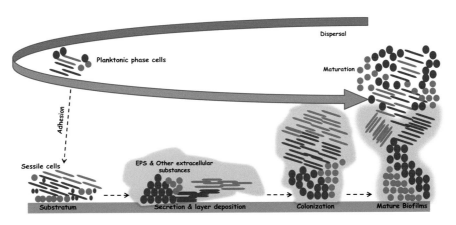

Fig. 1.3 Biofilm developments to a solid surface

(Pesci et al. 1999). The biofilm maturation and dispersion is the last stage in the biofilm development (Sauer et al. 2002).

To eradicate microbial biofilms, innovative complexes capable of combating biofilm development without any toxicity to the environment are developed. Biosurfactants are competent to impede with biofilm destruction, restraining microbial dealings with interfaces (Federle and Bassler 2003; Merk et al. 2005; Neu 1996; Sharma et al. 2015; Saharan et al. 2011; Rivardo et al. 2009). Biosurfactants have been shown to modify the physicochemical properties by changing the wettability properties and charge of the surface which leads to the reduction of the interaction of bacteria with the surface (Neu 1996; Banat et al. 2010). Conditioning of surface with biosurfactant downregulates the genes responsible for the biofilm formation film, which will modify the characteristic (wettability) of the surface (Neu 1996). New understandings into biofilm composition have now empowered researchers to design more effective biofilm dispersal approaches. Improvement of effective strategies, establishment on the onset of biofilm formation and their control is anticipated to be a foremost breakthrough in the anticipatory medicines.

Several reports showed a significant reduction in the initial biofilm deposition (Zeraik and Nitschke 2010; Rodrigues et al. 2006). The preconditioning of solid surfaces using biosurfactants may be a remarkable approach for combating the adhesion of food-borne pathogens to the solid surfaces. The significant activity confirmed by biosurfactants advocates that they could be deliberated as new tools in mounting strategies to avoid or interrupt microbial colonization of industrial surfaces used in food processing.

Future Prospects

As demonstrated by the rising figures of reports on the biosurfactants, there is an accumulative attention in the study of these biomolecules and their prospective applications. The demand for novel biosurfactants in the cosmetics, food, and pharmaceutical formulations, is progressively growing and the biosurfactants with effective and eco-friendly composition, impeccably meet this demand. Most of the biosurfactant-producing microorganisms are pathogenic and challenging to handle in commercial formulations. The progress of biosurfactant production from non-pathogenic microorganisms such as "Biosurfactant derived from LAB" is a prevailing task that is receiving augmented attention in direction to escape pathogenicity. Strategies making biosurfactant production cost effective includes optimized production cultural conditions and novel single step product recovery. Due to their self-assembly behavior, novel, and potential applications in nanoparticle synthesis are anticipated for biosurfactants. Detailed studies of their natural roles in microbial interactions, cell signaling, pathogenicity, and biofilm development could advocate significant future applications.

References

Abalos A, Pinazo A, Infante MR, Casals M, Garcia F, Manresa A (2001) Physicochemical and antimicrobial properties of new rhamnolipids produced by *Pseudomonas aeruginosa* AT10 from soybean oil refinery wastes. Langmuir 17(5):1367–1371

Abdallah FB, Chaieb K, Zmantar T, Kallel H, Bakhrouf A (2009) Adherence assays and slime production of *Vibrio alginolyticus* and *Vibrio parahaemolyticus*. Braz J Microbiol 40(2): 394–398

Adjonu R, Doran G, Torley P, Agboola S (2014) Whey protein peptides as components of nanoemulsions: a review of emulsifying and biological functionalities. J Food Eng 122:15–27

Agboola O, Maree J, and Mbaya R (2014) Characterization and performance of nanofiltration membranes. Environ chem Lett 12(2):241–255

Ahimou F, Jacques P, Deleu M (2000) Surfactin and iturin A effects on *Bacillus subtilis* surface hydrophobicity. Enzym Microbial Technol 27(10):749–754

Alvarez F, Castro M, Príncipe A, Borioli G, Fischer S, Mori G, Jofre E (2012) The plant-associated Bacillus amyloliquefaciens strains MEP218 and ARP23 capable of producing the cyclic lipopeptides iturin or surfactin and fengycin are effective in biocontrol of sclerotinia stem rot disease. J Appl Microbiol 112(1):159–174

Amézcua-Vega C, Poggi-Varaldo HM, Esparza-García F, Ríos-Leal E, Rodríguez-Vázquez R (2007) Effect of culture conditions on fatty acids composition of a biosurfactant produced by *Candida ingens* and changes of surface tension of culture media. Bioresour Technol 98(1):237–240

Arutchelvi JI, Bhaduri S, Uppara PV, Doble M (2008) Mannosylerythritol lipids: a review. J Ind Microbiol Biotechnol 35(12):1559–1570

Banat IM (1995) Biosurfactants production and possible uses in microbial enhanced oil recovery and oil pollution remediation: a review. Bioresour Technol 51(1):1–12

Banat IM, Makkar RS, Cameotra SS (2000) Potential commercial applications of microbial surfactants. Appl Microbiol Biotechnol 53(5):495–508

Banat IM, Samarah N, Murad M, Horne R, Banerjee S (1991) Biosurfactant production and use in oil tank clean-up. World J Microbiol Biotechnol 7(1):80–88

Banat IM et al (2010) Microbial biosurfactants production, applications and future potential. Appl Microbiol Biotechnol 87(2):427–444

Batista SB, Mounteer AH, Amorim FR, Totola MR (2006) Isolation and characterization of biosurfactant/bioemulsifier-producing bacteria from petroleum contaminated sites. Bioresour Technol 97(6):868–875

Benincasa M, Abalos A, Oliveira I, Manresa A (2004) Chemical structure, surface properties and biological activities of the biosurfactant produced by *Pseudomonas aeruginosa* LBI from soapstock. Antonie Van Leeuwenhoek 85(1):1–8

Bodour AA, Drees KP, Maier RM (2003) Distribution of biosurfactant-producing bacteria in undisturbed and contaminated arid southwestern soils. Appl Environ Microbiol 69(6):3280–3287

Bonmatin JM, Laprévote O, Peypoux F (2003) Diversity among microbial cyclic lipopeptides: iturins and surfactins. Activity-structure relationships to design new bioactive agents. Comb Chem High Throughput Screening 6(6):541–556

Bueno J (2014) Anti-biofilm drug susceptibility testing methods: looking for new strategies against resistance mechanism. J Microbial Biochem Technol 2014

Buijssen KJ, Harmsen HJ, van der Mei HC, Busscher HJ, van der Laan BF (2007) *Lactobacilli*: important in biofilm formation on voice prostheses. Otolaryngol Head Surg 137(3):505–507

Cameotra SS, Makkar RS (1998) Synthesis of biosurfactants in extreme conditions. Appl Microbiol Biotechnol 50(5):520–529

Cameotra SS, Makkar RS (2004) Recent applications of biosurfactants as biological and immunological molecules. Curr Opin Microbiol 7(3):262–266

Carrillo C, Teruel JA, Aranda FJ, Ortiz A (2003) Molecular mechanism of membrane permeabilization by the peptide antibiotic surfactin. Biochim Biophys Acta Biomembr 1611 (1):91–97

Chen XH et al (2009) Genome analysis of *Bacillus* amyloliquefaciens FZB42 reveals its potential for biocontrol of plant pathogens. J Biotechnol 140(1):27–37

Chmielewski RAN, Frank JF (2003) Biofilm formation and control in food processing facilities. Compr Rev Food Sci Food Saf 2(1):22–32

Das P, Mukherjee S, Sen R (2008) Improved bioavailability and biodegradation of a model polyaromatic hydrocarbon by a biosurfactant producing bacterium of marine origin. Chemosphere 72(9):1229–1234

Davies D (2003) Understanding biofilm resistance to antibacterial agents. Nat Rev Drug Discov 2 (2):114–122

Debode J, Maeyer KD, Perneel M, Pannecoucque J, Backer GD, Höfte M (2007) Biosurfactants are involved in the biological control of *Verticillium microsclerotia* by *Pseudomonas* spp. J Appl Microbiol 103(4):1184–1196

Deleu M, Paquot M, Nylander T (2008) Effect of fengycin, a lipopeptide produced by *Bacillus subtilis*, on model biomembranes. Biophys J 94(7):2667–2679

Desai JD, Banat IM (1997) Microbial production of surfactants and their commercial potential. Microbiol Mol Biol Rev 61(1):47–64

Dettenkofer M, Block C (2005) Hospital disinfection: efficacy and safety issues. Curr Opin Infect Dis 18(4):320–325

Donlan RM, Costerton JW (2002) Biofilms: survival mechanisms of clinically relevant microorganisms. Clin Microbiol Rev 15(2):167–193

Douillard FP, de Vos WM (2014) Functional genomics of lactic acid bacteria: from food to health. Microb Cell Fact 13:S8

Dunne WM (2002) Bacterial adhesion: seen any good biofilms lately? Clin Microbiology Rev 15 (2):155–166

Dusane DH, Zinjarde SS, Venugopalan VP, Mclean RJ, Weber MM, Rahman PK (2010) Quorum sensing: implications on rhamnolipid biosurfactant production. Biotechnol Genet Eng Rev 27 (1):159–184

El Aidy S, van den Bogert B, Kleerebezem M (2015) The small intestine microbiota, nutritional modulation and relevance for health. Curr Opin Biotechnol 32:14–20

Falagas ME, Makris GC (2009) Probiotic bacteria and biosurfactants for nosocomial infection control: a hypothesis. J Hosp Infect 71(4):301–306

Federle MJ, Bassler BL (2003) Interspecies communication in bacteria. J Clin Invest 112(9):1291–1299

Ferradji FZ, Mnif S, Badis A, Rebbani S, Fodil D, Eddouaouda K, Sayadi S (2014) Naphthalene and crude oil degradation by biosurfactant producing *Streptomyces* spp. isolated from Mitidja plain soil (North of Algeria). Int Biodeterior Biodegradation 86:300–308

Fracchia L, Banat IM, Martinotti MG, Cavallo M (2012) Biosurfactants and bioemulsifiers biomedical and related applications-present status and future potentials. INTECH Open Access Publisher

Fuentes MS et al (2014) Methoxychlor bioremediation by defined consortium of environmental *Streptomyces* strains. Int J Environ Sci Technol 11(4):1147–1156

Ghosh S, RingØ E, Selvam ADG, Rahiman KM, Sathyan N, John N, Hatha AAM (2014) Gut associated lactic acid bacteria isolated from the estuarine fish *Mugil cephalus*: molecular diversity and antibacterial activities against pathogens. Int J Aquac 4:1–11

Grover M, Nain L, Singh SB, Saxena AK (2010) Molecular and biochemical approaches for characterization of antifungal trait of a potent biocontrol agent *Bacillus subtilis* RP24. Curr Microbiol 60(2):99–106

Hall-Stoodley L, Costerton JW, Stoodley P (2004) Bacterial biofilms: from the natural environment to infectious diseases. Nat Rev Microbio 2(2):95–108

Hildebrand PD, Braun PG, McRae KB, Lu X (1998) Role of the biosurfactant viscosin in broccoli head rot caused by a pectolytic strain of *Pseudomonas fluorescens*. Can J Plant Pathol 20(3):296–303

Hua Z, Chen J, Lun S, Wang X (2003) Influence of biosurfactants produced by *Candida antarctica* on surface properties of microorganism and biodegradation of n-alkanes. Water Res 37 (17):4143–4150

Imamura, Y et al (2008) Fusarium and Candida albicans biofilms on soft contact lenses: model development, influence of lens type, and susceptibility to lens care solutions. Antimicrob Agents Chemother 52(1):171–182

Khire JM, Khan MI (1994) Microbially enhanced oil recovery (MEOR). Part 1. Importance and mechanism of MEOR. Enzym Microb Technol 16(2):170–172

Khopade A, Ren B, Liu XY, Mahadik K, Zhang L, Kokare C (2012) Production and characterization of biosurfactant from marine *Streptomyces* species B3. J Colloid Interface Sci 367(1):311–318

Kim PI, Ryu J, Kim YH, ChI YT (2010) Production of biosurfactant lipopeptides Iturin A, fengycin and surfactin A from *Bacillus subtilis* CMB32 for control of *Colletotrichum gloeosporioides*. J Microbiol Biotechnol 20(1):138–145

Kiran G, Seghal et al (2010) Optimization and characterization of a new lipopeptide biosurfactant produced by marine Brevibacterium aureum MSA13 in solid state culture. Bioresour Technol 101(7):2389–2396

Kotulova D, Slobodnikova, L (2010) Susceptibility of Staphylococcus aureus biofilms to vancomycin, gemtamicin and rifampin. Epidemiologie, mikrobiologie, imunologie: casopis Spolecnosti pro epidemiologii a mikrobiologii Ceske lekarske spolecnosti JE Purkyne 59(2):80–87

Kostka JE et al (2011) Hydrocarbon-degrading bacteria and the bacterial community response in Gulf of Mexico beach sands impacted by the Deepwater Horizon oil spill. Appl Environ Microbiol 77(22):7962–7974

Kryachko Y, Nathoo S, Lai P, Voordouw J, Prenner EJ, Voordouw G (2013) Prospects for using native and recombinant rhamnolipid producers for microbially enhanced oil recovery. Int Biodeterior Biodegradation 81:133–140

Kulakovskaya TV, Golubev WI, Tomashevskaya MA, Kulakovskaya EV, Shashkov AS, Grachev AA, Nifantiev NE (2010) Production of antifungal cellobiose lipids by *Trichosporon porosum*. Mycopathologia 169(2):117–123

Kulakovskaya T, Shashkov A, Kulakovskaya E, Golubev W, Zinin A, Tsvetkov Y, Nifantiev N (2009) Extracellular cellobiose lipid from yeast and their analogues: structures and fungicidal activities. J Oleo Sci 58(3):133–140

Kumar CG, Anand SK (1998) Significance of microbial biofilms in food industry: a review. Int J Food Microbiol 42(1):9–27

Kumar AS, Mody K, Jha B (2007) Evaluation of biosurfactant/bioemulsifier production by a marine bacterium. Bull Environ Contam Toxicol 79(6):617–621

Kuyukina MS, Ivshina IB, Makarov SO, Litvinenko LV, Cunningham CJ, Philp JC (2005) Effect of biosurfactants on crude oil desorption and mobilization in a soil system. Environ Int 31 (2):155–161

Ławniczak Ł, Marecik R, Chrzanowski Ł (2013) Contributions of biosurfactants to natural or induced bioremediation. Appl Microbiol Biotechnol 97(6):2327–2339

Litzler PY et al (2007) Biofilm formation on pyrolytic carbon heart valves: influence of surface free energy, roughness, and bacterial species. J Thorac Cardiovasc Surg 134(4):1025–1032

Lu JR, Zhao XB, Yaseen M (2007) Biomimetic amphiphiles: biosurfactants. Curr Opin Colloid Interface Sci 12(2):60–67

Maneerat S (2005) Production of biosurfactants using substrates from renewable-resources. Songklanakarin J. Sci. Technol 27(3):675–683

Matsuyama T, Nakagawa Y (1996) Surface-active exolipids: analysis of absolute chemical structures and biological functions. J Microbiol Methods 25(2):165–175

Matsuyama T, Tanikawa T, Nakagawa Y (2011) Serrawettins and other surfactants produced by Serratia. In: Soberón-Chávez G (ed) Biosurfactants. Springer, Berlin Heidelberg, pp 93–120

Merk K, Borelli C, Korting HC (2005) *Lactobacilli*–bacteria–host interactions with special regard to the urogenital tract. Int J Med Microbiol 295(1):9–18

Miller RM, Zhang, Y (1997) Measurement of biosurfactant-enhanced solubilization and biodegradation of hydrocarbons. In: Sheehan D (ed) Bioremediation protocols, Humana Press, pp 59–66

Mimee B et al (2005) Antifungal activity of flocculosin, a novel glycolipid isolated from *Pseudozyma flocculosa*. Antimicrob Agents Chemother 49(4):1597–1599

Mohammadipour M, Mousivand M, Salehi Jouzani G, Abbasalizadeh S (2009) Molecular and biochemical characterization of Iranian surfactin-producing *Bacillus subtilis* isolates and evaluation of their biocontrol potential against *Aspergillus flavus* and *Colletotrichum gloeosporioides*. Can J Microbiol 55(4):395–404

Morikawa M (2006) Beneficial biofilm formation by industrial bacteria *Bacillus subtilis* and related species. J Biosci Bioeng 101(1):1–8

Morita T, Fukuoka T, Imura T, Kitamoto D (2013) Production of mannosylerythritol lipids and their application in cosmetics. Appl Microbiol Biotechnol 97(11):4691–4700

Mulligan CN (2009) Recent advances in the environmental applications of biosurfactants. Curr Opin Colloid Interface Sci 14(5):372–378

Naruse N et al (1990) Pumilacidin, a complex of new antiviral antibiotics. Production, isolation, chemical properties, structure and biological activity. J Antibiot 43(3):267–280

Neu TR (1996) Significance of bacterial surface-active compounds in interaction of bacteria with interfaces. Microbiol Rev 60(1):151

Nilsson RE, Ross T, Bowman JP (2011) Variability in biofilm production by *Listeria monocytogenes* correlated to strain origin and growth conditions. Int J Food Microbiol 150(1):14–24

Nitschke M, Costa SG, Haddad R, Gonçalves G, Lireny A, Eberlin MN, Contiero J (2005) Oil wastes as unconventional substrates for rhamnolipid biosurfactant production by *Pseudomonas aeruginosa* LBI. Biotechnol Prog 21(5):1562–1566

Nitschke M, Costa SGVAO (2007) Biosurfactants in food industry. Trends Food Sci Technol 18(5):252–259

Nitschke M, Costa SG, Contiero J (2010) Structure and applications of a rhamnolipid surfactant produced in soybean oil waste. Appl Biochem Biotechnol 160(7):2066–2074

Oliveira FJS, Vazquez L, De Campos NP, De Franca FP (2009) Production of rhamnolipids by a *Pseudomonas alcaligenes* strain. Process Biochem 44(4):383–389

Ongena M, Jacques P (2008) *Bacillus* lipopeptides: versatile weapons for plant disease biocontrol. Trends Microbiol 16(3):115–125

O'Toole GA., Kolter R (1998) Flagellar and twitching motility are necessary for Pseudomonas aeruginosa biofilm development. Mol Microbiol 30(2):295–304

Ouwehand AC, Salminen S, Isolauri E (2002) Probiotics: an overview of beneficial effects. Antonie Van Leeuwenhoek 82(1–4):279–289

Pamp SJ, Tolker-Nielsen T (2007) Multiple roles of biosurfactants in structural biofilm development by *Pseudomonas aeruginosa*. J Bacteriol 189(6):2531–2539

Parsek MR, Singh PK (2003) Bacterial biofilms: an emerging link to disease pathogenesis. Annu Rev Microbiol 57(1):677–701

Perfumo A, Smyth TJP, Marchant R, Banat IM (2010) Production and roles of biosurfactants and bioemulsifiers in accessing hydrophobic substrates. In: Timmis KN (ed) Handbook of hydrocarbon and lipid microbiology. Springer, Berlin Heidelberg, pp 1501–1512

Pesci EC, Iglewski BH (1999) Quorum sensing in Pseudomonas aeruginosa. Cell-cell signaling in bacteria. American Society for Microbiology, Washington, pp 147–155

Petrelli D et al (2006) Analysis of different genetic traits and their association with biofilm formation in *Staphylococcus epidermidis* isolates from central venous catheter infections. Eur J Clin Microbiol Infect Dis 25(12):773–781

Peypoux F, Bonmatin JM, Wallach J (1999) Recent trends in the biochemistry of surfactin. Appl Microbiol Biotechnol 51(5):553–563

Rahman PK, Gakpe E (2008) Production, characterisation and applications of biosurfactants –review. Biotechnology 7(2):360–370

Rahman KSM, Banat IM, Thahira J, Thayumanavan T, Lakshmanaperumalsamy P (2002) Bioremediation of gasoline contaminated soil by a bacterial consortium amended with poultry litter, coir pith and rhamnolipid biosurfactant. Bioresour Technol 81(1):25–32

Rahman KSM, Rahman TJ, Kourkoutas Y, Petsas I, Marchant R, Banat IM (2003) Enhanced bioremediation of n-alkane in petroleum sludge using bacterial consortium amended with rhamnolipid and micronutrients. Bioresour Technol 90(2):159–168

Rivardo F, Turner RJ, Allegrone G, Ceri H, Martinotti MG (2009) Anti-adhesion activity of two biosurfactants produced by *Bacillus* spp. prevents biofilm formation of human bacterial pathogens. Appl Microbiol Biotechnol 83(3):541–553

Rodrigues LR, Teixeira JA (2010) Biomedical and therapeutic applications of biosurfactants. In: Soberón-Chávez G (ed) Biosurfactants, Springer, New York, pp 75–87

Rodrigues L, Banat IM, Teixeira J, Oliveira R (2006) Biosurfactants: potential applications in medicine. J Antimicrob Chemother 57(4):609–618

Rodrigues L, Van der Mei HC, Teixeira J, Oliveira R (2004) Influence of biosurfactants from probiotic bacteria on formation of biofilms on voice prostheses. Appl Environ Microbiol 70(7):4408–4410

Ron EZ, Rosenberg E (2002) Biosurfactants and oil bioremediation. Curr Opin Biotechnol 13(3):249–252

Rosenberg E, Ron EZ (1997) Bioemulsans: microbial polymeric emulsifiers. Curr Opin Biotechnol 8(3):313–316

Rosenberg E, Ron EZ (1999) High-and low-molecular-mass microbial surfactants. Appl Microbiol Biotechnol 52(2):154–162

Sachdev DP, Cameotra SS (2013) Biosurfactants in agriculture. Appl Microbiol Biotechnol 97(3):1005–1016

Saharan BS, Sahu RK, Sharma D (2011) A review on biosurfactants: fermentation, current developments and perspectives. Genet Eng Biotechnol J 1:1–14

Sauer K et al (2002) Pseudomonas aeruginosa displays multiple phenotypes during development as a biofilm. J Bacteriol 184(4):1140–1154

Sekhon KK, Khanna S, Cameotra SS (2012) Biosurfactant production and potential correlation with esterase activity. J Pet Environ Biotechnol 3(7):133

Seydlová G, Svobodová J (2008) Review of surfactin chemical properties and the potential biomedical applications. Cent Eur J Med 3(2):123–133

Shah V, Jurjevic M, Badia D (2007) Utilization of restaurant waste oil as a precursor for sophorolipid production. Biotechnol Prog 23(2):512–515

Sharma D, Singh Saharan B (2014) Simultaneous Production of biosurfactants and bacteriocins by probiotic *Lactobacillus casei* MRTL3. Int J Microbiol 2014

Sharma D et al (2015) Isolation and functional characterization of novel biosurfactant produced by Enterococcus faecium. SpringerPlus 4(1):1–14

Singh P, Cameotra SS (2004) Potential applications of microbial surfactants in biomedical sciences. Trends in Biotechnol 22(3):142–146

Snook ME, Mitchell T, Hinton DM, Bacon CW (2009) Isolation and characterization of Leu7-surfactin from the endophytic bacterium *Bacillus mojavensis* RRC 101, a biocontrol agent for *Fusarium verticillioides*. J Agric Food Chem 57(10):4287–4292

Stoodley P et al (2002) Biofilms as complex differentiated communities. Annu Rev Microbiol 56(1):187–209

Stickler DJ (2008) Bacterial biofilms in patients with indwelling urinary catheters. Nat Clin Pract Urol 5(11):598–608

Tran H, Ficke A, Asiimwe T, Höfte M, Raaijmakers JM (2007) Role of the cyclic lipopeptide massetolide A in biological control of *Phytophthora infestans* and in colonization of tomato plants by *Pseudomonas fluorescens*. New Phytol 175(4):731–742

Tran H, Kruijt M, Raaijmakers JM (2008) Diversity and activity of biosurfactant-producing Pseudomonas in the rhizosphere of black pepper in Vietnam. J Appl Microbiol 104(3):839–851

Toren A et al (2002) Solubilization of polyaromatic hydrocarbons by recombinant bioemulsifier AlnA. Appl microbiol biotechnol 59(4–5):580–584

Van Bogaert IN, Saerens K, De Muynck C, Develter D, Soetaert W, Vandamme EJ (2007) Microbial production and application of sophorolipids. Appl Microbiol Biotechnol 76(1):23–34

Van Hamme JD, Singh A, Ward OP (2006) Physiological aspects: Part 1 in a series of papers devoted to surfactants in microbiology and biotechnology. Biotechnol Adv 24(6):604–620

Vollenbroich D, Özel M, Vater J, Kamp RM, Pauli G (1997) Mechanism of inactivation of enveloped viruses by the biosurfactant surfactin from *Bacillus subtilis*. Biologicals 25(3):289–297

Walia NK, Cameotra SS (2015) Lipopeptides: Biosynthesis and Applications. J Microb Biochem Technol 7:103–107

Zeraik AE, Nitschke M (2010) Biosurfactants as agents to reduce adhesion of pathogenic bacteria to polystyrene surfaces: effect of temperature and hydrophobicity. Curr Microbiol 61(6):554–559

Chapter 2
Biosurfactants of Probiotic Lactic Acid Bacteria

Abstract According to the definition of Food and Agriculture Organization of the Unites Nations and World Health Organization (FAO/WHO) "Probiotics are the live microbial preparation, when consumed, confer the health benefits to the consumer." Lactic acid bacteria are recognized to produce various antimicrobial compounds such as bacteriocin, biosurfactants organic acids, carbon peroxide, diacetyl, low molecular weight antimicrobial substances, and hydrogen peroxide, which prevent the growth of potential pathogens. The use and possible application of biosurfactants in the biomedicals had increased in past decade. Their antimicrobial properties make them appropriate molecules for combating many pathogens and as therapeutic agents. Furthermore, their application as antibiofilm agents against commonly known pathogens specifies their effectiveness as appropriate anti-adhesive coating agents for biomedical insertional equipment. The present chapter covers all the biomedical aspects of biosurfactants of lactic acid bacteria in medical and therapeutic perceptions.

Keywords Biosurfactant · Antibiofilm · Antimicrobial · Probiotics and biomedical surfaces

Introduction

Lactic acid bacteria (LAB) include widespread genera comprising an extensive number of species involved in fermentation of dairy products. Along with the potential in dairy fermentation, LAB are extensively found as the part of animal and human intestinal commensal flora (Vaughan et al. 2005). The roles of LAB have been recognized in maintaining homeostasis within active ecosystems such as the urinary tract and gastrointestinal tract to control colonization by various pathogens (Boris and Barbes 2000). The research on microorganism of probiotics origin has gained attention in current biomedical applications worldwide. Research observations advises that probiotic microorganisms may have a vital role in declining the occurrence of antibiotic-related diarrhoea, prevention of vaginal candidiasis, improved

© The Author(s) 2016
D. Sharma et al., *Biosurfactants of Lactic Acid Bacteria*,
SpringerBriefs in Microbiology, DOI 10.1007/978-3-319-26215-4_2

immunological defense responses, and urinary tract infections (Falagas et al. 2006; Falagas and Makris 2009). Mainly biosurfactant-producing microorganisms are pathogenic in nature and difficult to handle in industrial formulations (Sharma et al. 2015; Toribio et al. 2010; Saharan and Nehra 2011). The application of biosurfactants derived from pathogenic microorganism in industrial formulations is predominantly objectionable precisely in the food processing, cosmetics formulations, and pharmaceuticals (Saharan et al. 2011; Banat and Desai 1997).

Probiotics are: "Preparation of Live microorganisms which when administered in adequate amounts confer a health benefit on the host" (FAO/WHO 2002). Probiotics have been reported to have encouraging effects on the maintenance of human and animal health (Gupta and Garg 2009). Awareness in probiotics research and formulations have extended pronounced significance due to the rise in antimicrobial resistance worldwide. Probiotics microorganisms are recognized to produce various antimicrobial compounds such as bacteriocin, biosurfactants organic acids, carbon peroxide, diacetyl, low molecular weight antimicrobial substances, and hydrogen peroxide, which prevent the growth of potential pathogens (Pascual et al. 2008; Merk et al. 2005; Ceresa et al. 2015) (Fig. 2.1). Various studies reported the prospective of LABS as biosurfactant producers and their potential role in public health and food processing (Velraeds et al. 1996; Heinemann et al. 2000; Rodrigues et al. 2004; Servin 2004; Rodrigues et al. 2006, Falagas and Makris 2009; Thavasi et al. 2011; Gudiña et al. 2011; Rodríguez-Pazo et al. 2013; Moldes et al. 2013; Sharma et al. 2014, 2015) (Table 2.1).

Furthermore, probiotics have long been acknowledged also for the potential to restrict the adhesion and development of biofilms of pathogens to the various biological (epithelial cells of urogenital and intestinal tracts) and inanimate (food

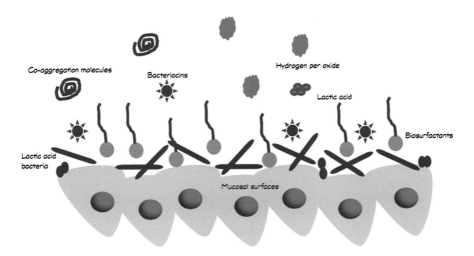

Fig. 2.1 Production of various compounds by LAB on mucosal membrane

Table 2.1 Strains of LAB reported for biosurfactant production and their application

S. No.	LAB Strain	Source of strain	Application	Study
1.	*Lactobacillus* spp.	Urogenital tract of healthy women	Antiadhesive against the uropathogenic *Enterococcus faecalis*	Velraeds et al. (1996)
2.	*Streptococci thermophilus*	Heat exchanger plate of pasteurizer	Antiadhesive activity against *Candida* sp.	Busscher and Van der Mei (1997)
3.	*Streptococci mitis*	Human oral cavity	Inhibition of *Streptococci mutans*	Van Hoogmoed et al. (2006)
4.	*Lactobacillus fermentum* RC-14	Urogenital isolate of a healthy woman	Inhibits adhesion of *Enterococcus faecalis* 1131	Heinemann et al. (2000)
5.	*Lactococcus lactis* 53	–	Antibiofilm activity	Rodrigues et al. (2004)
6.	*Lactobacillus casei* CECT 525, *Lactobacillus rhamnosus* CECT 288, *Lactobacillus pentosus* CECT 4023 *and Lactobacillus coryniformis* subsp. *torquens* CECT 25600	Spanish culture collection center	Kinetics of biosurfactants production	Rodrigues et al. (2006)
7.	*Streptococcus thermophiles* A	NIZO	Antimicrobial and antiadhesive properties	Rodrigues et al. (2006)
8.	*Lactobacillus pentosus*	–	Biosurfactant production	Rivera et al. (2007)
9.	*Lactobacillus pentosus*	CECT 4023T	Biosurfactant production	Moldes et al. (2013)
10.	*Lactobacillus acidophillus*	–	Antiadhesive activity against *Staphylococcus aureus*	Walencka et al. (2008)
11.	*Lactobacillus acidophillus*	CECT 419	Biosurfactant production	Portilla et al. (2008)
12.	*Lactococcus lactis*	Curd sample	Antibacterial activity against the multidrug resistant pathogens	Sarvanakumari and Mani (2010)
13.	*Lactobacillus paracasei*	Portuguese dairy industry	Antimicrobial and antiadhesive properties	Gudiña et al. (2010)
14.	*Lactococcus lactis*	CECT 4434	Simultaneous extraction of biosurfactant and bacteriocin	Rodríguez et al. (2010)

(continued)

Table 2.1 (continued)

S. No.	LAB Strain	Source of strain	Application	Study
15.	*Lactococcus paracasei* subsp. *Paracasei* A20	Portuguese dairy industry	Antimicrobial and Antiadhesive properties	Gudiña et al. (2010)
16.	*Lactobacillus delbrueckii* sbusp. *delbruckii*	DSMZ	Biosurfactant inhibition against *Candida albicans*	Fracchia et al. (2010)
17.	*Lactobacillus delbrueckii*	–	Biosurfactant production and structural characterization	Thavasi et al. (2011)
18.	*Lactobacillus acidophilus*	DSM	Effect on GTFB and GTFC expression level	Tahmourespour et al. (2011)
19.	*Lactobacillus fermentii* and *Lactobacillus rhamnosus*	CCCIFM, Polnad	Antiadhesive properties against the *Klebsiella pneumonia*, *Pseudomonas aeruginosa* and *E. coli*	Brzozowski et al. (2011)
20.	*Lactobacillus* spp.	Yogurt, Cheese and Silage	Antagonistic activity	Kermanshahi et al. (2012)
21.	*Lactobacillus plantarum*	–	Structural characterization	Sauvageau et al. (2012)
22.	*Lactobacillus reutri*	DSM	Effect of biosurfactant on gene expression of essential adhesion genes (gtfB, gtfC and ftf) of *Streptococcus mutans*	Salehi et al. (2014)
23.	*Lactobacillus* spp.	Pendidam	Antibacterial activity	Augustin et al. (2014)
24.	*Lactobacillus pentosus* and *Lactobacillus plantarum* co-culture	CECT 4023 CECT 221	Production and antimicrobial activity	Rodriguez-Pazo et al. (2013)
25.	*Lactobacillus* spp.	Egyptian dairy product	Antimicrobial activity	Gomaa (2013)
26.	*Lactobacillus pentosus*	–	Foaming agents	Vecino et al. (2013)
27.	*Lactobacillus plantarum* CFR2194	Kanjika (rice based ayurvedic fermented product)	Antimicrobial and Antiadhesive activity	Madhu and Prapulla (2014)
28.	*Lactobacillus pentosus*	–	Fatty acid characterization	Vecino et al. (2014)
29.	*Lactobacillus pentosus*	–	Structural properties	Vecino et al. (2014)

(continued)

Table 2.1 (continued)

S. No.	LAB Strain	Source of strain	Application	Study
30.	*Lactobacillus brevis*	Fresh cabbage	Antifungal activity	Ceresa et al. (2015)
31.	*Lactobacillus jensenii* and *Lactobacillus rhamnosus*	American type culture collection (ATCC)	Biofilm dispersal and antimicrobial	Sambanthamoorthy et al. (2014)

processing area, biomedical surfaces, and food equipment) (Reid et al. 2001; Saharan et al. 2011; Sharma et al. 2015). LAB interfere with pathogen colonization by different mechanisms. Biosurfactant production is one of their mechanisms to prevent the colonization (Santos et al. 2013). Probiotic organisms probably interfere the adhesion by the release of biosurfactant molecules (Sharma et al. 2015; Gudiña et al. 2010; Rodrigues et al. 2006). Production of lipopeptides by *Bacillus* probiotics prevent the growth of pathogens existing in the gastrointestinal tract (Hong et al. 2005). Likewise, antagonism with other pathogens for adherence to the epithelial cells along with biosurfactants secretion is a well recognized mechanisms used by lactic acid bacteria to obstruct vaginal pathogens (Cribby et al. 2009; Falagas et al. 2007).

LAB-Derived Biosurfactants in Biomedical

Biofilm in microbial system is a community attached to either biotic or abiotic planes rooted by an extra-polymeric matrix for endurance under unfavorable conditions (Donlan and Costerton 2002). Microbial biofilms were first described in 1943 (Zobell 1943), but the problem is still persisting in an extensive kind of extents, particularly in the food (Veran et al., 2010) and biomedical sector (Sihorkar and Vyas 2001). Often frequently, microbial biofilm structures are a regular challenge confronted by the food processing sector. The predominance of biofilms is a substantial problem in food formulation and food processing (Murphy et al. 2006; Gandhi and Chikindas 2007). In food industries, a range of microorganisms colonize the food contact surfaces and form biofilm microbial communities. Once established, microbial biofilms are a substantial source of contamination of food products.

Biofilms sheltering multi-antibiotic-resistant microorganisms are predominantly established at astonishing levels on hospital surfaces and leads to the risk of nosocomial infection transmission. Biofilms in hospital environments are frequently associated with plastic medical tubing of various equipment. The incidence of multidrug resistant pathogens being sheltered inside these biofilms are problematic in biomedical surfaces (Vandecandelaere and Coenye 2015).

Numerous reports have pointed out that probiotic microorganism's derived biosurfactants may combat the growth of hospital acquired pathogens on inanimate surfaces and biomedical surfaces (Sharma et al. 2015; Rodrigues et al. 2004, 2006; Walencka et al. 2008). Biosurfactants derived from the LAB has a valuable application as antiadhesive agents to combat the colonization of pathogenic microorganisms. Application of biosurfactant to any surface modifies its hydrophobicity, interfering in microbial adhesion. Contribution of biosurfactants in microbial adhesion has been extensively defined, and establishes a potential approach to reduce microbial adhesion by pathogens, not merely in the biomedical applications, but also in other capacities, such as the food processing industry (Falagas and Makris 2009). Various observations have been made regarding the antiadhesive nature of biosurfactants derived from the LAB. Reduction of pathogen colonization has been reported in glass (Velraeds et al. 1996), silicon, rubber prostheses (Busscher and Van der Mei 1997; Velraeds et al. 1998; Van Hoogmoed et al. 2006; Rodrigues et al. 2004, 2006) metal (Meylheuc et al. 2006), and other inanimate surfaces (Heinemann et al. 2000; Gudiña et al. 2010; Fracchia et al. 2010).

Falagas and Makris (2009) studied in vitro tests on the significant role of LAB-derived biosurfactants in the inhibition of fungal and bacterial colonization on various surfaces such as urinary catheters and other voice prostheses or surgical implants made of silicone rubber (Rodrigues et al. 2004, 2006). The application of biosurfactant mainly falls in two categories; majorly pre-coating of the surfaces with microbial surfactants, or directly added to biosurfactant-producing LAB strains to inspect biofilm development. Additional exciting application region that is gaining increased attention relates to probiotics use in avoiding oral infections (Meurman 2005; Meurman and Stamatova 2007; Vujic et al. 2013; Toiviainen 2015). Biosurfactant derived from *Streptococcus mitis* inhibited the adhesion of tooth decaying bacteria *Streptococcus sobrinus* and *Streptococcus mutans* to enamel, whereas *S. mitis* biosurfactant was capable to constrain the adhesion of *S. sobrinus* to salivary pellicles (Van Hoogmoed et al. 2004). Novel biosurfactant molecules derived from LAB are reported from different dairy products for anti-biofilm properties (Walencka et al. 2008; Sharma and Saharan 2014; Sharma et al. 2014, 2015). Furthermore, biosurfactant accumulation to redeveloped, settled biofilms enhanced their dispersal and changed the biofilm morphology.

Biosurfactant produced by *Lactococcus lactis* (Xylolipid), isolated from a fermented dairy preparation exhibited significant antibacterial activity for clinical pathogens (Saravanakumari and Mani 2010). Considering their significance for human and animal health with acknowledged safety, LAB symbolizes a nontoxic and effective interference for pathogen control. Biosurfactant isolated from probiotics could be applied to biomedical equipment, such as silicone tubes and catheters, to combat microbial colonization of these surfaces by hospital-acquired pathogens (Falagas and Makris 2009).

It was usually reported that probiotic LAB mainly, *Streptococcus thermophilus* and *Lactobacillus* spp. strains derived surfactants antagonize the growth of certain pathogens like *Staphylococcus aureus*, *Streptococcus* spp., *Enterococcus faecalis*, *Candida albicans* (Van Hoogmoed et al. 2004). Rodrigues et al. (2004) confirmed that the biosurfactant derived from the *Lactococcus lactis* 53 was competent to constrain the adhesion of various pathogens to silicone tubing. Biosurfactant from probiotic strains significantly reduced the pathogenic microorganism's populations on voice prostheses (Rodrigues et al. 2004). Velraeds et al. (1996) also reported the control of biofilm developed by enteric pathogen with biosurfactant derived from *Lactobacillus* strain and later demonstrated that the biosurfactant triggered significant dose-related control of the preliminary deposition rate of *E. coli* adherent on hydrophobic and hydrophilic layers.

LAB-Derived Biosurfactants as Antimicrobials

The use and effective application of microbial surfactants in the biomedical sector has increased during the past couple of years. Biosurfactant derived from various microorganisms have been reported for antimicrobial properties (Sharma et al. 2015; Díaz De Rienzo et al. 2015; Cameotra and Makkar 2004). Recently, biosurfactants derived from the LAB have been specified to display antimicrobial properties. Various compounds produced by LAB have application in the production of newer generation antimicrobials (Reid et al. 2001; Rodrigues et al. 2004). Gram-positive bacteria are more profound against the biosurfactants than gram-negative bacteria, which were moderately inhibited. The biological properties of biosurfactant depend on the molecular structure of cell. In broad-spectrum, they influence the permeability of cellular plasma membranes.

The mechanism of biosurfactant antimicrobial activity still remains unclear, but certain hypotheses are proposed which are sustained by specific evidence displaying the loss of membrane integrity.

- Antimicrobial activity of the microbial surfactants is an outcome of the adhesion property of these surface active agents to the cell surfaces instigating decline of cell membrane integrity leads to subsequent collapse of the nutrition cycle (Inès and Dhouha 2015).
- The fatty acid moieties of biosurfactants inserting into the cell membrane instigating a proliferation of membrane size and ultra-structural changes (Gomaa 2013).
- Biosurfactants are also able to form pores and disrupt the plasma membrane (Inès and Dhouha 2015) (Fig. 2.2).
- Addition of the smaller acyl tails of the biosurfactant into the plasma membrane triggering disruptions of the plasma membrane, countenancing the plasma membrane to lift away from the cytoplasmic matter (Desai and Banat 1997).

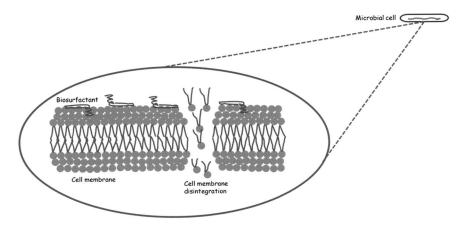

Fig. 2.2 Mode of action of biosurfactants

- Disruption of cell membranes from side to side buildup of intra membranous elements in the cells increasing the electrical conductance of the plasma membrane
- Increasing cell membrane sponginess through the disruption of the plasma membrane phospholipids (Carrillo et al. 2003; Sotirova et al. 2009).
- Biosurfactant exposure to the cells changed the fatty acids contents in the cell membrane due to the disturbance in plasma membrane permeability. Biosurfactant directly interact with the lipids, which trigger inhibition of the membrane-confined enzyme and outflow of intracellular cytoplasmic components. In other study, Cameotra and Makkar (2004) described that the antimicrobial potential of biosurfactant could intrude the plasma membrane structure while interacted with phospholipids and other membrane proteins.
- Treatment of *S. aureus* cells with the biosurfactants resulted in the reduction of lipids content and also had a relation with the disruption of proteins present in the plasma membrane. The phenomenon may probably be accredited to the property of increasing membrane protein initiating the conformational changes of lipid and protein molecules.

Several biosurfactants are recognized to have therapeutic claims as antifungal, antibacterial, and antiviral complexes (Inès and Dhouha 2015). Antimicrobial potential of biosurfactants makes them appropriate complexes for applications in contending to numerous diseases and act as a therapeutic agents (Table 2.2). For illustration, the antimicrobial potential of two biosurfactants moieties derived from probiotic bacteria, *Lactococcus lactis* 53 and *Streptococcus thermophilus* A, against various pathogenic bacteria and yeast strains colonizing voice prostheses were assessed (Rodrigues et al. 2004).

Biosurfactant producing lactobacilli with certain health benefits have been frequently isolated from the urinary tract and intestinal tracts (Reid et al. 2001). There

Table 2.2 Antimicrobial properties of biosurfactants derived from LAB

S. No.	Strain	Antimicrobial activity	References
1.	*Streptococcus thermophiles* A	Antimicrobial activity against the *Candida tropicalis*	Rodrigues et al. (2006)
2.	*Lactobacillus casei*	Antimicrobial activity against *Staphylococcus aureus, Bacillus subtilis,* and *Micrococcus roseus*	Gołek et al. (2009)
3.	*Lactococcus lactis*	Antimicrobial activity of biosurfactant (Xylolipids) against the multidrug resistant *Staphylococcus aureus* and *E. coli*	Sarvanakumari and Mani (2010)
4.	*Lactobacillus paracasei*	Growth inhibition of *E. coli, S. agalactiae,* and *S. pyogenes* with a concentration of 25 mg/ml	Gudiña et al. (2010)
5.	*Lactobacillus paracasei* A20	Antimicrobial activity against various gram-positive and gram-negative microorganisms at various concentration ranging from 3.12 mg/ml to 50 mg/ml	Gudiña et al. (2010)
6.	*Lactobacillus casei* MRTL3	Antimicrobial activity against *Staphylococcus aureus* ATCC 6538P, *S. epidermidis* ATCC 12228, *Bacillus cereus* ATCC 11770, *Listeria monocytogenes* MTCC 657, and *L. innocua* ATCC 33090, *Shigella flexneri* ATCC 9199, *Salmonella typhi* MTCC 733	Sharma and Saharan (2014)

is a high incidence of pathogens associated with the surgical implants mainly caused by the *S. aureus*. Biosurfactants produced by the probiotic strain of *Lactobacillus fermentum* RC-14 reduced the population and adherence of *S. aureus* to the surgical implants (Gan et al. 2002). Rodrigues et al. (2006) demonstrated biosurfactant production by *S. thermophiles* inhibit the growth of *Candida tropicalis* at the concentration of 2.5 g/l. The crude biosurfactant showed significant antimicrobial properties against the *E. coli, S. agalactiae,* and *S. pyogenes* with a concentration of 25 mg/ml (Gudiña et al. 2010). Biosurfactant derived from the *Lactobacillus paracasei* A20 showed antimicrobial activity against various gram-positive and gram-negative microorganisms at various concentrations ranging from 3.12 mg/ml to 50 mg/ml. The biosurfactant derived from *Lactobacillus casei* MRTL3 isolated from raw milk showed significant antimicrobial properties against various pathogens, including *Staphylococcus aureus, S. epidermidis, Bacillus cereus, Listeria monocytogenes* and *L. innocua, Shigella flexneri, Salmonella typhi,* and *Pseudomonas aeruginosa* (Sharma and Saharan 2014).

Conculusion

Biosurfactant derived from the genus *Lactobacillus* are of biomedical and food interests. Biosurfactants derived from LAB are advantageous as antibacterial and antifungal agents, and they also have the potential for use as antibiofilm agents in biomedical and food processing. Encouraging substitutes to conventional antibiotics with less toxicity and broad spectrum may be exploited for their biomedical importance. Moreover, biosurfactants of lactic acid bacteria have the potential to be used as antibiofilm biological coatings for biomedical equipment for decreasing nosocomial pathogens. Biosurfactant of lactic acid bacteria may also be integrated into probiotic formulations to combat urinary tract infections. Although there is immense potential of biosurfactants of lactic acid bacteria in the biomedical and food processing, their application still remains inadequate, conceivably due to high production cost, toxicity, and inadequate evidence on their structural details. Further research investigations on their toxicity and structural attributes are desired to authenticate the use of LAB-derived biosurfactants in various biomedical and food formulations.

References

Augustin M, Majesté PM, Hippolyte MT (2014) Effect of manufacturing practices on the microbiological quality of fermented milk (Pendidam) of some localities of Ngaoundere (Cameroon). Int J Curr Microbiol App Sci 3(11):71–81

Boris S, Barbés C (2000) Role played by *lactobacilli* in controlling the population of vaginal pathogens. Microbes Infect 2(5):543–546

Brzozowski B, Bednarski W, Golek P (2011) The adhesive capability of two *Lactobacillus* strains and physicochemical properties of their synthesized biosurfactants. Food Technol Biotechnol 49(2):177

Busscher HJ, Van der Mei HC (1997) Physico-chemical interactions in initial microbial adhesion and relevance for biofilm formation. Adv Dent Res 11(1):24–32

Cameotra SS, Makkar RS (2004) Recent applications of biosurfactants as biological and immunological molecules. Curr Opin Microbiol 7(3):262–266

Carrillo C, Teruel JA, Aranda FJ, Ortiz A (2003) Molecular mechanism of membrane permeabilization by the peptide antibiotic surfactin. Biochim Biophys Acta (BBA)-Biomembranes 1611(1): 91–97

Ceresa C, Tessarolo F, Caola I, Nollo G, Cavallo M, Rinaldi M, Fracchia L (2015) Inhibition of Candida albicans adhesion on medical-grade silicone by a Lactobacillus-derived biosurfactant. J Appl Microbiol 118(5):1116–1125

Cribby S, Taylor M, Reid G (2009) Vaginal microbiota and the use of probiotics. Interdisciplinary perspectives on infectious diseases 2008

De Rienzo MAD, Banat IM, Dolman B, Winterburn J, Martin PJ (2015) Sophorolipid biosurfactants: possible uses as antibacterial and antibiofilm agent. New Biotechnol 32 (6):720–726

Desai JD, Banat IM (1997) Microbial production of surfactants and their commercial potential. Microbiol Mol Biol Rev 61(1): 47–64

Donlan RM, Costerton JW (2002) Biofilms: survival mechanisms of clinically relevant microorganisms. Clin Microbiol Rev 15(2):167–193

Falagas ME, Makris GC (2009) Probiotic bacteria and biosurfactants for nosocomial infection control: a hypothesis. J Hosp Infect 71(4):301–306

Falagas ME, Betsi GI, Athanasiou S (2006) Probiotics for prevention of recurrent vulvovaginal candidiasis: a review. J Antimicrob Chemother 58(2):266–272

Falagas ME, Betsi GI, Athanasiou S (2007) Probiotics for the treatment of women with bacterial vaginosis. Clin Microbiol Infect 13(7):657–664

Fracchia L, Cavallo M, Allegrone G, Martinotti MG (2010) A *Lactobacillus*-derived biosurfactant inhibits biofilm formation of human pathogenic *Candida albicans* biofilm producers. Appl Microbiol Biotechnol 2:827–837

Gandhi M, Chikindas ML (2007) *Listeria*: a foodborne pathogen that knows how to survive. Int J Food Microbiol 113(1):1–15

Gan BS, Kim J, Reid G, Cadieux P, Howard JC (2002) Lactobacillus fermentum RC-14 inhibits *Staphylococcus aureus* infection of surgical implants in rats. J Infect Dis 185(9):1369–1372

Gołek P, Bednarski W, Brzozowski B, Dziuba B (2009) The obtaining and properties of biosurfactants synthesized by bacteria of the genus*Lactobacillus*. Ann Microbiol 59(1):119–126

Gomaa EZ (2013) Antimicrobial activity of a biosurfactant produced by *Bacillus* licheniformis strain M104 grown on whey. Braz Arch Biol Technol 56(2):259–268

Gudiña EJ, Teixeira JA, Rodrigues LR (2010) Isolation and functional characterization of a biosurfactant produced by *Lactobacillus paracasei*. Colloids and Surfaces B: Biointerfaces 76(1): 298–304

Gudiña EJ, Rocha V, Teixeira JA, Rodrigues LR (2010) Antimicrobial and antiadhesive properties of a biosurfactant isolated from *Lactobacillus paracasei* ssp. *paracasei* A20. Lett Appl Microbiol 50(4):419–424

Gupta V, Garg R (2009) Probiotics. Indian J Med Microbiol 27(3):202

Heinemann C, van Hylckama Vlieg JE, Janssen DB, Busscher HJ, van der Mei HC, Reid G (2000) Purification and characterization of a surface-binding protein from *Lactobacillus fermentum* RC-14 that inhibits adhesion of *Enterococcus faecalis* 1131. FEMS Microbiol Lett 190(1):177–180

Hong HA, Duc LH, Cutting SM (2005) The use of bacterial spore formers as probiotics. FEMS microbiology reviews 29(4):813–835

Inès M, Dhouha G (2015) Lipopeptide surfactants: Production, recovery and pore forming capacity. Peptides 71:100–112

Kermanshahi RK, Peymanfar S (2012) Isolation and identification of lactobacilli from cheese, yoghurt and silage by 16S rDNA gene and study of bacteriocin and biosurfactant production. Jundishapur j Microbiol 5(4):528–532

Madhu AN, Prapulla SG (2014) Evaluation and functional characterization of a biosurfactant produced by *Lactobacillus plantarum* CFR 2194. Appl. Biochem. Biotechnol 172(4):1777–1789

Merk K, Borelli C, Korting HC (2005) *Lactobacilli*–bacteria–host interactions with special regard to the urogenital tract. Int J Med Microbiol 295(1):9–18

Meurman JH (2005) Probiotics: do they have a role in oral medicine and dentistry? Eur J Oral Sci 113(3):188–196

Meurman JH, Stamatova I (2007) Probiotics: contributions to oral health. Oral Dis 13(5):443–451

Meylheuc T, Methivier C, Renault M, Herry JM, Pradier CM, Bellon-Fontaine MN (2006) Adsorption on stainless steel surfaces of biosurfactants produced by gram-negative and gram-positive bacteria: consequence on the bioadhesive behavior of *Listeria monocytogenes*. Colloids Surf B 52(2):128–137

Moldes AB, Paradelo R, Vecino X, Cruz JM, Gudiña E, Rodrigues L, Barral MT (2013) Partial characterization of biosurfactant from *Lactobacillus pentosus* and comparison with sodium dodecyl sulphate for the bioremediation of hydrocarbon contaminated soil. BioMed Res Int 961842:6 pp

Murphy C, Carroll C, Jordan KN (2006) Environmental survival mechanisms of the food borne pathogen *Campylobacter jejuni*. J Appl Microbiol 100(4):623–632

Pascual LM, Daniele MB, Giordano W, Pájaro MC, Barberis IL (2008) Purification and partial characterization of novel bacteriocin L23 produced by *Lactobacillus fermentum* L23. Curr Microbiol 56(4):397–402

Portilla-Rivera O, Torrado A, Domínguez JM, Moldes AB (2008) Stability and emulsifying capacity of biosurfactants obtained from lignocellulosic sources using *Lactobacillus pentosus*. J Agric Food Chem 56(17):8074–8080

Reid G, Howard J, Gan BS (2001) Can bacterial interference prevent infection? Trends Microbiol 9(9):424–428

Reid G, Zalai C, Gardiner G (2001) Urogenital lactobacilli probiotics, reliability, and regulatory issues. J Dairy Sci 84: E164–E169

Rivera OMP, Moldes AB, Torrado AM, Domínguez JM (2007) Lactic acid and biosurfactants production from hydrolyzed distilled grape marc. Process Biochem 42(6):1010–1020

Rodrigues L, Van der Mei H, Teixeira JA, Oliveira R (2004) Biosurfactant from *Lactococcus lactis* 53 inhibits microbial adhesion on silicone rubber. Appl Microbiol Biotechnol 66(3):306–311

Rodrigues L, Banat IM, Teixeira J, Oliveira R (2006) Biosurfactants: potential applications in medicine. J Antimicrob Chemother 57(4):609–618.

Rodríguez N, Salgado JM, Cortés S, Domínguez JM (2010). Alternatives for biosurfactants and bacteriocins extraction from *Lactococcus lactis* cultures produced under different pH conditions. Lett Appl Microbiol 51(2):226–233

Rodríguez-Pazo N, Salgado JM, Cortés-Diéguez S, Domínguez JM (2013) Biotechnological production of phenyllactic acid and biosurfactants from trimming vine shoot hydrolyzates by microbial coculture fermentation. Appl Biochem Biotechnol 169(7):2175–2188

Rodríguez-Pazo N, Vázquez-Araújo L, Pérez-Rodríguez N, Cortés-Diéguez S, Domínguez JM (2013) Cell-free supernatants obtained from fermentation of cheese whey hydrolyzates and phenylpyruvic acid by L*actobacillus plantarum* as a source of antimicrobial compounds, bacteriocins, and natural aromas. Appl Biochem Biotechnol 171(4):1042–1060

Saharan BS, Nehra V (2011) Plant growth promoting rhizobacteria: a critical review. Life Sci Med Res 21:1–30

Saharan BS, Sahu RK, Sharma D (2011) A review on biosurfactants: fermentation, current developments and perspectives. Genet Eng Biotechnol J 2011(1):1–14

Salehi R, Savabi O, Kazemi M (2014) Effects of Lactobacillus reuteri-derived biosurfactant on the gene expression profile of essential adhesion genes (gtfB, gtfC and ftf) of *Streptococcus mutans*. Adv Biomed Res 3

Sambanthamoorthy K, Feng X, Patel R, Patel S, Paranavitana C (2014) Antimicrobial and antibiofilm potential of biosurfactants isolated from *lactobacilli* against multi-drug-resistant pathogens. BMC Microbiol 14(1):1

Santos DK, Rufino RD, Luna JM, Santos VA, Salgueiro AA, Sarubbo LA (2013) Synthesis and evaluation of biosurfactant produced by Candida lipolytica using animal fat and corn steep liquor. J Petrol Sci Eng 105:43–50

Saravanakumari P, Mani K (2010) Structural characterization of a novel xylolipid biosurfactant from *Lactococcus lactis* and analysis of antibacterial activity against multi-drug resistant pathogens. Bioresour Technol 101(22):8851–8854

Sauvageau J, Ryan J, Lagutin K, Sims IM, Stocker BL, Timmer MS (2012) Isolation and structural characterisation of the major glycolipids from *Lactobacillus plantarum*. Carbohydr Res 357:151–156

Servin AL (2004) Antagonistic activities of *lactobacilli* and bifidobacteria against microbial pathogens. FEMS Microbiol Rev 28(4):405–440

Sharma D, Saharan BS, Chauhan N, Procha S, Lal S (2015) Isolation and functional characterization of novel biosurfactant produced by *Enterococcus faecium*. SpringerPlus 4(1):1–14

Sharma D, Singh Saharan B (2014) Simultaneous Production of biosurfactants and bacteriocins by probiotic *Lactobacillus casei* MRTL3. Int J Microbiol 2014

Sharma D, Saharan BS, Chauhan N, Bansal A, Procha S (2014) Production and structural characterization of *Lactobacillus helveticus* derived biosurfactant. Sci World J 2014

Sihorkar V, Vyas SP (2001) Biofilm consortia on biomedical and biological surfaces: delivery and targeting strategies. Pharm Res 18(9):1247–1254

Sotirova A, Spasova D, Vasileva-Tonkova E, Galabova D (2009). Effects of rhamnolipid-biosurfactant on cell surface of Pseudomonas aeruginosa. Microbiol Res 164(3): 297–303

Tahmourespour A, Salehi R, Kermanshahi RK, Eslami G (2011) The anti-biofouling effect of Lactobacillus fermentum-derived biosurfactant against *Streptococcus mutans*. Biofouling 27 (4):385–392

Thavasi R, Jayalakshmi S, Banat IM (2011) Application of biosurfactant produced from peanut oil cake by *Lactobacillus delbrueckii* in biodegradation of crude oil. Bioresour Technol 102 (3):3366–3372

Toiviainen A (2015 September) Probiotics and oral health: In vitro and clinical studies. Annales Universitatis Turkuensis, Sarja–Ser. D, Medica-Odontologica

Toribio J, Escalante AE, Soberón-Chávez G (2010) Rhamnolipids: Production in bacteria other than *Pseudomonas aeruginosa*. Eur J Lipid Sci Technol 112(10):1082–1087

van Hoogmoed CG, van der Mei HC, Busscher HJ (2004) The influence of biosurfactants released by S. mitis BMS on the adhesion of pioneer strains and cariogenic bacteria. Biofouling 20 (6):261–267

Van Hoogmoed CG, Dijkstra RJB, Van der Mei HC, Busscher HJ (2006) Influence of biosurfactant on interactive forces between mutans streptococci and enamel measured by atomic force microscopy. J Dent Res 85(1):54–58

Vandecandelaere I, Coenye T (2015) Microbial composition and antibiotic resistance of biofilms recovered from endotracheal tubes of mechanically ventilated patients. Biofilm-based Healthcare-associated Infections. Springer International Publishing. 137–155

Vaughan EE, Heilig HG, Ben-Amor K, De Vos WM (2005) Diversity, vitality and activities of intestinal lactic acid bacteria and bifidobacteria assessed by molecular approaches. FEMS Microbiol Rev 29(3):477–490

Velraeds MM, Van de Belt-Gritter B, Van der Mei HC, Reid G, Busscher HJ (1998) Interference in initial adhesion of uropathogenic bacteria and yeasts to silicone rubber by a *Lactobacillus acidophilus* biosurfactant. J Med Microbiol 47(12):1081–1085

Velraeds MM, Van der Mei HC, Reid G, Busscher HJ (1996) Inhibition of initial adhesion of uropathogenic *Enterococcus faecalis* by biosurfactants from *Lactobacillus* isolates. Appl Environ Microbiol 62(6):1958–1963

Vecino X, Devesa-Rey R, Cruz JM, Moldes AB (2013). Evaluation of biosurfactant obtained from *Lactobacillus pentosus* as foaming agent in froth flotation. J Environ Manage 128:655–660

Vecino X, Devesa-Rey R, Moldes AB, Cruz JM (2014) Formulation of an alginate-vineyard pruning waste composite as a new eco-friendly adsorbent to remove micronutrients from agroindustrial effluents. Chemosphere 111:24–31

VERAN—TISSOIRES S, GERINO M (2010) Imaging biofilm in porous media using X-ray computed microtomography

Vujic G, Knez AJ, Stefanovic VD, Vrbanovic VK (2013) Efficacy of orally applied probiotic capsules for bacterial vaginosis and other vaginal infections: a double-blind, randomized, placebo-controlled study. Eur J Obstetr Gynecol Reprod Biol 168(1):75–79

Walencka E, Różalska S, Sadowska B, Różalska B (2008) The influence of *Lactobacillus acidophilus*-derived surfactants on staphylococcal adhesion and biofilm formation. Folia Microbiol 53(1):61–66

Zobell CE (1943) The effect of solid surfaces upon bacterial activity. J Bacteriol 46(1):39

Chapter 3
Properties of Biosurfactants

Abstract The growing environmental worry about synthetic chemical surfactants initiates responsiveness to microbial-derived biosurfactants fundamentally due to their low toxicity, stability, and biodegradable nature. Biosurfactants are primarily used in bioremediation of hydrocarbon pollutants; though, they show possible applications in various sectors of food processing industries. Related with emulsion forming and stabilization, antibiofilm and antimicrobial properties are some advantageous properties of biosurfactants, which could be recognized in food processing and formulation. Prospective applications of biosurfactants in food formulations and the use of agro-industrial substrates for their cost-effective production are discussed.

Keywords Surface tension · Interfacial tension · Critical micelle concentration · Toxicity

Introduction

Biosurfactants are produced mainly by microorganisms including bacteria and fungi (including yeasts), and have potential surface active properties. The solubility, surface tension reduction, critical micelle concentration together with detergent properties, wetting and foaming potential will make a biosurfactant suitable for particular applications (Saharan et al. 2011; Banat et al. 2010; Myers 2005). Similar properties are shown by chemically synthesized surfactants which are produced from petrochemical or oleochemical substrates (Desai and Banat 1997) and these chemical surfactants have been broadly developed for large-scale industrial applications.

Biosurfactants are regarded as less toxic and environmentally safe and thus various kinds of microbial surfactants can be commercially produced for broad applications in pharmaceutical formulations, cosmetics preparation, and food processing. Commonly, surfactants have various well-known functional characteristics, and exploited in numerous industrial sectors.

© The Author(s) 2016
D. Sharma et al., *Biosurfactants of Lactic Acid Bacteria*,
SpringerBriefs in Microbiology, DOI 10.1007/978-3-319-26215-4_3

- Surface tension
- Critical micelle concentration
- Stability

Surface Tension

The organized forces among liquid molecules are accountable for the phenomenon known as the surface tension (ST). It creates a surface "film" that makes it additionally challenging to transfer an object from the surface rather transfer it when it is entirely dipped. The identical state also applies at the interface of the two fluids that do not mix with each other like oil in water and it is known as interfacial tension (IFT) (Fig. 3.1). Typically, unit of measurement for surface and interfacial tension is mN/m (which is comparable to dynes/cm). The potential of a biosurfactant is evaluated by its ability to decrease the surface tension of production broth. An effective biosurfactant can reduce the surface tension of water from 72.0 to 35.0 mN/m (Mulligan 2005). As per the described standards, a minimal reduction of surface tension by 8 mN/m is to be considered as an effective biosurfactant producing microorganism (Van der Vegt et al. 1991).

The biosurfactant derived from *Enterococcus faecium* MRTL9, can substantially decrease the surface tension of phosphate buffer saline (PBS) extract from 72 to 40.2 mN/m. Numerous LAB strains were described as an efficient biosurfactant producer on the basis of their potential to decrease the surface tension of production broth. Goek et al. (2009) also reported a substantial reduction in surface tension from 72 to 35.5 mN/m while culturing different strains of *L. casei* and *Bifidobacterium*. LAB generally do not reduce the surface tension as comparable

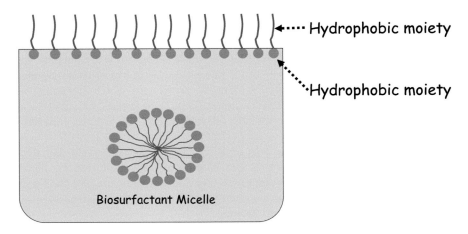

Fig. 3.1 Behavior of biosurfactant in liquid interfaces

with *Pseudomonas aeruginosa* and *Bacillus subtilis* but certain thermophillic strains of LAB like *Streptococcus thermophilus* A and mesophillic *Lactococcus lactis* 53 decrease surface tension to 36.0–37.0 from 72.0 mN/m (Rodrigues et al. 2004).

Cell-bound biosurfactant derived from the *L. acidophilus* cells recovered from fermentation broth reduced surface tension 18–24.5 mN/m units from 72.0 mN/m in vine-trimming hydrolysate medium (Portilla-Riviera et al. 2008). Maximal surface reduction was found with cell-bound biosurfactant as compared to the extracted biosurfactant in the supernatant. While comparing the reduction in surface tension, maximal reduction in surface tension of PBS was found, i.e. (24.5 units) as compared to the strain grown in hemicellulose hydrolysate medium (18 units).

Intracellular and extracellular biosurfactants derived from *L. pentosus* were observed by using hemicellulose sugars-based fermentative medium obtained from the distinct lignocellulosic residues as carbon source. *L. pentosus* cells were harvested from the broth and the biosurfactant was extracted in PBS. Highest reduction in surface tension was observed while the strain is grown on hemicellulose carbon source, i.e., 21.4 units (Bustos et al. 2007). Various authors reported the significant surface tension reduction while working with different LAB strains, especially reduction in surface tension was found maximal when the intracellular biosurfactant was extracted with gentle stirring in PBS (Table 3.1). In a case, Gudiña et al. (2010) reported that biosurfactant extracted from *L. paracasei* reduced the surface tension value from 72.0 to 41.8 mN/m. Velreads et al. (1996) reported maximal reduction in surface tension by *L. fermentum* RC14 strain, i.e., 39 with a cmc value of 1.0 mg/ml. The maximal reduction in surface tension was reported in the stationary phase of growth.

Among all the studied LAB strains for biosurfactant production, it was found that the amount of cell-bound biosurfactant is found to be maximal compared to the excreted biosurfactant isolated from the cell suspension. On the other hand, hyper-biosurfactant-producing strains of *Pseudomonas* and *Bacillus* sp. produce mainly excreted biosurfactants. Gudiña et al. (2013) established the fact that biosurfactant production from LAB was found to be growth associated and there was a parallel relationship among carbon source utilization, biomass production, and surface tension reduction.

Critical Micelle Concentration

The CMC (Critical Micelle Concentration) is the amount of surfactant molecule in bulk stage, beyond which aggregates of surface active agents are formed. The CMC is a significant distinctive property of surfactants for their industrial and environmental applications. Generally, molecules have two different constituents with differing attraction for the solutes. The component of the molecule that has empathy for polar solutes, like water, is assumed to be hydrophilic. The component of the molecule with empathy for nonpolar solutes, like hydrocarbons, is assumed to be

Table 3.1 Surface tension reduction by biosurfactants derived from different LAB

S. No.	Strain	Surface tension (mN/m)	References
1.	*L. casei*	53.0	Rodrigues et al. (2006)
2.	*L. rhamnosus*	51.5	Rodrigues et al. (2006)
3.	*L. pentosus*	50.5	Rodrigues et al. (2006)
4.	*L. coryniformis*	55.0	Rodrigues et al. (2006)
5.	*Streptococcus thermophilus* A	37.0	Rodrigues et al. (2006)
6.	*Lactobacillus pentosus*	56.6	Rivera et al. (2007)
7.	*Lactobacillus pentosus*	51.0	Bustos et al. (2007)
8.	*Lactobacillus pentosus*	53.9	Portilla-Rivera et al. (2008)
9.	*L. casei* subsp. rhamnosus	45.0	Velraeds et al. (1996)
10.	*Lactobacillus acidophilus* CECT-419	47.5	Portilla-Riviera et al. (2008)
11.	*L. casei* 1825	53.9	Goek et al. (2009)
12.	*L. fermentum* 126	53.7	Goek et al. (2009)
13.	*L. plantarum* 91	57.3	Goek et al. 2009
14.	*L. brevis* T16	58.5	Goek et al. (2009)
15.	*L. brevis* 37	55.2	Goek et al. (2009)
16.	*L. paracasei*	41.9	Gudiña et al. (2010)
17.	*L. lactis*	41.1	Rodríguez et al. (2010)
18.	*Lactobacillus* spp.	45.4	Fracchia et al. (2010)
19.	*L. paracasei* ssp. *paracasei* A20	47.5	Gudiña et al. (2013)
20.	*L. plantarum*	54.2	Rodriguez-Pazo et al. (2013)
21.	*L. pentosus*	53.0	Vecino et al. (2013)
22.	*L. plantarum* CFR2194	44.3	Madhu and Prapulla (2014)
23.	*L. fermentum* RC14	39.0	Valreads et al. (1996)
24.	*Entercoccus faecium* MRTL9	40.2	Sharma et al. (2015)
25.	*L. helveticus* MRTL 91	39.5	Sharma et al. (2014)

hydrophobic. And the same molecule with hydrophilic and hydrophobic moieties is known to be amphiphilic. Amphiphilic molecules demonstrate different conduct when act together with water. An amphiphilic molecule can dispose itself on the surface of aqueous phase so that the polar moiety interacts with the aqueous phase and the nonpolar moiety is detained above the surface (in the air/nonpolar liquid). Such compounds are known as "surface active agents" or surfactants. Microbial cells (bacteria and yeast) are reported ubiquitously for producing this kind of surfactant in their ecological niche and known as biosurfactants (Saharan et al. 2011). Potent surface active agents have low CMC amount, i.e., very less concentration of surfactant is essential to reduce the surface tension (George and Jayachandran 2009). The CMC observations for microbial surfactants generally range between

Table 3.2 CMC of biosurfactants derived from various strains of LAB

S. No.	Strain	Critical micelle concentration (CMC)	References
1.	*L. acidophilus*	1.0 mg/mL	Velraeds et al. (1996)
2.	*S. thermophiles* A	20 g/L	Rodrigues et al. (2006)
3.	*L. pentosus*	4.8 mg/L	Rivera et al. (2007)
4.	*L. fermentum* RC14	1.0 mg/L	Velraeds et al. (1996)
5.	*L. fermentum* B54	2.0 mg/L	Velraeds et al. (1996)
6.	*S. thermophiles* B	2.0 mg/L	Busscher et al. (1994)
7.	*Lactococcus lactis* 53	14.0 mg/L	Rodrigues et al. (2006)
8.	*Lactobacillus* spp. CV8LAC	106 µg/mL	Fracchia et al. (2010)
9.	*Entercococcus faecium* MRTL9	2.25 mg/L	Sharma et al. (2015)
10.	*L. pentosus*	2.2 mg/L	Vecino Bell et al. (2012)
11.	*L. plantarum*	5.6 mg/L	Rodriguez-Pazo et al. (2013)
12.	*L. plantarum*	2.0 mg/L	Rodriguez-Pazo et al. (2013)
13.	*L. pentosus*	1.3 mg/L	Rodriguez-Pazo et al. (2013)
14.	*L. pantarum* CFR2194	6 g/L	Madhu and Prapulla (2014)
15.	*L. pentosus*	10 g/L	Vecino et al. (2014)

5–100 mg/L (Table 3.2). At low concentrations, biosurfactants reside on the surface of the water. As the surface gets crowded with biosurfactant molecules, surplus biosurfactant molecules assemble in the micelle formation (Fig. 3.2).

The biosurfactant molecules combined at CMC due to fragile hydrophobic and van der Waals contacts and the micelles form to entangle hydrophobic substances and in their emulsification with water. The CMC of biosurfactants is usually 10–40 folds lower than chemically synthesized surfactants. Biosurfactant at low CMC can attain higher emulsification potential in diverse industrial and biotechnological practices (Desai and Banat 1997; Muthusamy et al. 2008).

Biosurfactants derived from various LAB range between 1.0 mg/L and 20 g/L as reported from various studies previously. Although the biosurfactants produced by the various strain of LAB were not as efficient as compared to the biosurfactant isolated from the *Bacillus* and *Pseudomonas* genus (Mulligan et al. 2001), the biosurfactant produced by *L. fermentum* RC14 was 1.0 mg/mL while organism is grown under stationary phase; surface tension reached approximately 39 mN/m. Biosurfactant release by lactobacilli was maximal under the stationary phase of growth (Valreads et al. 1996). Rivera et al. (2007) has reported that the biosurfactant isolated from *L. pentosus* showed a CMC value 2.6 mg/L which is the

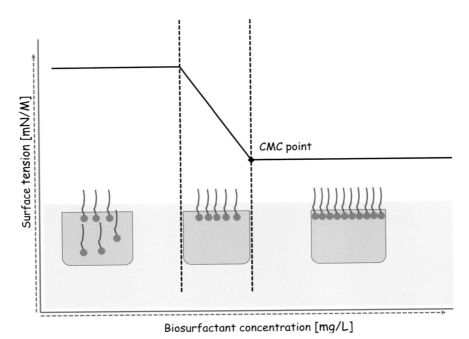

Fig. 3.2 Critical micelle concentration

lowest concentration of biosurfactant that allows to gain the highest surface tension reduction. Above the concentration, recognized as the CMC, it is not conceivable to endure reduction in surface tension due to the circumstance that the biosurfactant subordinates willingly to form micelles and vesicles. It has been reported that the CMC was attained for the biosurfactants extracts acquired when grape marc and hazelnut and walnut coat hemicellulosic extracts were engaged as carbon source. It was observed that the CMC corresponded to the biosurfactant derived from walnut shell hemicellulose hydrolyzates. Biosurfactant derived from distilled grape marc medium had a CMC parallel to the biosurfactants produced with vine shoot trimming hydrolyzates medium, whereas hazelnut shells-derived hemicellulose produced the surfactant with the higher CMC (Portilla-Riviera et al. 2008).

The CMC of the crude biosurfactant derived from the probiotic strain *L. paracasei* isolated from the Portuguese dairy industry was found to be 2.5 mg/ml with surface tension 41.8 mN/m (Gudiña et al. 2013). Those surfactants can decrease the surface tension of water or PBS around 36–39 mN/m and even their CMC values ranged from 1.0–20 mg/mL. When compared with the chemical surfactant, it had a surface tension reduction and CMC values very close to the surface tension reduction by SDS (Sodium dodecyl sulfate), i.e., 37.0 mN/m and 1.8 mg/L, respectively. Biosurfactant produced by the *Lactobacillus* spp. CV8LAC, which inhibits the biofilm formation by human pathogen, was found to be 106 μg/mL with a reduction in surface tension from 70.92 to 45.4 mN/m (Fracchia et al. 2010).

The optimal release of biosurfactant from *Lactobacillus* spp. CV8LAC is observed in the mid-exponential growth phase. Velraeds et al. (1996) reported that the biosurfactants released in the stationary phase from *Lactobacillus* spp. are higher in amount than in the mid-exponential phase and had a better distinct CMC. Similarly, Gudiña et al. (2010) extracted biosurfactants from the stationary phase of the production with a low CMC value. Simultaneous fermentation of hemicellulosic hydrolysate medium and solid state fermentation of cellulosic fraction by co-culture of *L. plantarum* and *L. pentosus* reduced the surface tension with a CMC value of 2.6 mg/L. CMC value was found to be lower in case of submerged fermentation as compared to the solid state fermentation (Rodriguez-Pazo et al. 2013). The role of biosurfactant derived from the *L. pentosus* as a foaming agent has been studied by Vecino et al. (2013) and it was found that the CMC is around 2 mg/L. The biosurfactant of *L. pentosus* could be diluted almost sixfold with decrease in the surface tension of the liquid. Biosurfactant produced by the *L. plantarum* CFR 2194 is structurally composed of protein and polysaccharides moieties, i.e., a glycoprotein with a CMC of 6 g/L. Information about the CMC is quiet vital when using surfactants in industrial formulations. Surface tension does not decrease additionally above the CMC, or in various preparations the CMC stipulates the restraining concentration for significant use.

Emulsification

An emulsion is a combination of two or more solutions that are usually immiscible (Fig. 3.3). Emulsions are portion of a more common class of two-phase systems of substance called colloids. Illustrations of emulsions comprises of milk and mayonnaise in food preparations. The word "emulsion" originates from the Latin language word for "milk." According to the surface tension theory, emulsification takes place by decreasing the interfacial tension among two phases.

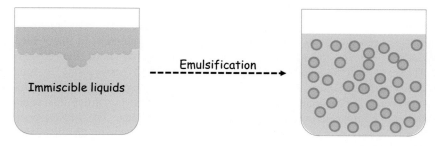

Fig. 3.3 Biosurfactant-mediated emulsification

Bioemulsifiers are biological in nature and are produced either by the microbial cells or extracellularly produced by emulsifying compounds (Neu and Karl 1990). The external appendages of various bacterial cells are surrounded by extracellularly produced polysaccharides, i.e., EPS occur both as distinct capsules or copious slime layer.

About 220 patents are on the data relating these biosurfactants like Emulsan (Liu et al. 2010) have been broadly considered and extended industrial applications, mainly in food processing, cosmetics, and pharmaceuticals formulations. The significant emulsification potential is vital for biosurfactants to be promising in distinct environmental and commercial applications (Saharan et al. 2011).

Together from their apparent role as surface active agents that reduce surface and interfacial tension, biosurfactants have various other significance in food processing. Biosurfactants are used in improving the texture of food formulations, agglomeration of fat drops and increasing the shelf life of starch-based products, and improving the uniformity and texture of fat-based food preparations.

Microbial surfactants may also be considered for their emulsifying potential. Emulsification potentials are conferred in expressions of emulsifying stabilities, emulsifying capacity, and emulsification activity of biosurfactants-based preparations. The emulsification stability is generally recorded in relations to the period that an extent of oil separating from an emulsion reliant on temperature, gravitational force, and the amount of oil (Lima et al. 2011).

The biosurfactants are also efficiently used in bakery products for controlling consistency in cake icing, delaying staling, and can be consumed in fat stabilization, and antispattering substances (Nitschke and Costa 2007; Kachholz and Schlingmann 1987). High-molecular-mass microbial surfactants are broadly improved emulsifying agents than low-molecular mass surfactants. Liposan are kind of biosurfactant that had displayed less surface tension reduction properties but were effective to emulsify edible oils (Cirigliano and Carman 1985). The study also advised the use of rhamnolipids biosurfactants to expand the properties of butter cream and frozen confectionery formulations. Determination of emulsifying potential of biosurfactants is overall associated to hydrocarbons because of their effective applications in environmental issues. Fewer efforts have been completed on the use of biosurfactants in emulsion establishment in food industry (Nitschke and Pastore 2006).

The use of LAB-derived biosurfactants is motivating as these microorganisms are generally recognized as safe (GRAS) for human consumption in many food formulations, on the other aspect, biosurfactants derived from the devious *P. aeruginosa*, still face certain resistance regarding their applicability as food ingredients or formulations. *L pentosus* grown on lignocellulosic hydrolysate produces a surface active agent with significant emulsification properties that could expedite the bioremediation of hydrocarbon contaminations. The biosurfactant derived from the *L. pentosus* produces biosurfactant when grown on distilled grape marc hydrolysate and gave 50 % (v/v) relative emulsification volume and were stable after 72 h of incubation against kerosene oil. The value of emulsification volume compared to the commercial surfactant was found to be higher, i.e., 14.1 %

for gasoline and 27.2 % for kerosene. But contrary to commercial surfactant, biosurfactant derived from *L. pentosus* gave higher emulsification volume. Furthermore, biosurfactant produced using distilled grape marc showed stable emulsification, i.e., 99 % as compared to the biosurfactants produced using hazelnut shell hydrolysate, i.e., 97 %. Additionally, emulsification with the chemically synthesized surfactant, i.e., SDS showed only 87.7 % (Portilla-Riviera et al. 2008). The consumption of low-cost fermentative medium as substrate for production of bioemulsifiers is the possible strategy for the large-scale production and industrial applications of the bioemulsifiers produced by the microorganisms. Thavasi et al. (2011) observed the emulsification potential of the biosurfactant produced by the *L. delbruckii*. Biosurfactant produced by the *L. delbruckii* using peanut cake waste as a substrate gave emulsification activity of 1.85 ± 0.14 as compared to the emulsification activity showed by chemical surfactant Triton-X-100, i.e., 1.93 ± 0.08. Although the emulsification activity produced by the *L. delbruckii* was low as compared to the Triton-X-100, the advantages over the chemical surfactant were less toxicity, biodegradability, ecologically safe, and could be advantageous for the bioremediation applications.

The order of emulsification activity against various hydrocarbons was: motor lubricant oil > crude oil > peanut oil > kerosene > diesel > xylene > anthracene > naphthalene. Biosurfactant derived from the *L. delbruckii* using peanut oil cakes as a potent substrate showed its effectiveness to be used in bioremediation approaches. Even though the biosurfactant used in emulsification of hydrocarbons need not to be of high purity, their applicability in bioremediation of hydrocarbons is highly advantageous (Thavasi et al. 2011). Portilla-Rivera et al. (2008) showed that biosurfactants obtained from the *L. pentosus* can stabilize water and gas oil emulsion in contrast with SDS, surfactin and Tween 80 and it was found that biosurfactants derived from the *L. pentosus* provided enhanced percentages of emulsions (16 %). On the contrary, surfactin and Tween 80 gave 9 and 0 % emulsification volume, respectively, after 48 h. Though SDS was observed to have superior emulsion percentage, i.e., 50 %. Effect of temperature on emulsification activity has been evaluated with biosurfactant obtained from the *Lactobacillus* spp. And it has been concluded that the biosurfactant maintains its emulsification potential almost unaffected by the temperature ranges between 25 and 100 °C after incubation for 48 h.

The emulsification index of the biosurfactant derived from *L. plantarum* CFR 2194 (1 mg/mL) was assayed against various hydrocarbons such as xylene, coconut oil, kerosene, hexane, heptane, and sunflower oil. In the middle of the hydrocarbons, heptane observed a maximum emulsification index value of 38.2, trailed by xylene (16.22), kerosene (15.2), and hexane (13.6). Emulsification index was considerably significant against different vegetable oils such as coconut oil (37.9) and sunflower oil (19.43), which can have broader applications for food processing (Madhu et al. 2014). It is perceived that vegetables oil were proficiently emulsified in contrast to kerosene and xylene. Emulsions formed with coconut and sunflower oils were additionally stable than with hydrocarbons. In the related reports, Moldes et al. (2013) and Vecino Bello et al. (2012) observed that microbial surfactants derived from *L. pentosus*, revealed to stabilize gasoline/water emulsions subsequently 24 h of incubation.

Vecino et al. (2014) observed the emulsions as oil-in water or water-in oil emulsion, stabilized by the cell-bound bisourfactant derived from the *L. pentosus* and polysorbate 20 against rosemary oil. The droplets of emulsion were rapidly mixed in water, but on the other side, when the emulsion was added to rosemary oil they persisted insolubilized form. Thus, the emulsion can be mentioned as an oil in water emulsion. The dimension of the droplets has a significant consequence on the constancy of the emulsion and optical properties like lightness and color, rheology and likewise its creaminess. The biosurfactant derived from the *L. pentosus* was capable to stabilize the spread droplets of rosemary oil in water; however, polysorbate 20 produced very uneven emulsions. The application of biosurfactants in food-grade oils and bioremediation depends highly upon the emulsification activity of the biosurfactant. Biosurfactant with superior emulsification activity and emulsification stability could be the agents of the future food emulsions and various other environmental applications.

Stability

The applicability of biosurfactant is highly dependent on their stability at various environmental conditions such as extreme temperatures and pH ranges (Sharma et al. 2015). For instance, biosurfactant used to emulsify vegetables oil in food processing should be stable at different temperature ranges for their effective action. Subsequently, biosurfactants used as formulation ingredient in various biomedical preparations like oral suspension, ointment base, and other superficial application drugs should be stable at different pH changes (Sharma et al. 2014). The stability of biosurfactant at different pH and temperature ranges would be anticipated for their application in various commercial and environmental applications. Distinct biosurfactant has been reported from various strains of lactic acid bacteria. Mainly the effect of different environmental conditions such as temperature, pH and salinity evaluated on the basis of their effect of surface tension and emulsification potential of the biosurfactants. Sharma et al. (2015) reported the stability of biosurfactant on different pH and temperature obtained from the *Enterococcus faecium* MRTL9 (Figs. 3.4 and 3.5). In their study, biosurfactant retained their activity, i.e., reduction in surface tension and emulsification activity over different pH range started from 6.0 to 10.0 and observed negligible deviance in surface tension and emulsification values. Biosurfactant was found to be stable over different pH shifts. Subsequently, biosurfactant obtained from the *E. faecium* was found to be stable even after treatment at 0–120 °C for 30 min. The effect of various pH and incubation temperatures on biosurfactant derived from the *E. faecium* at high temperature on surface tension and emulsifying potential were insignificant.

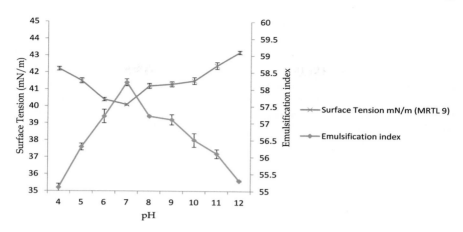

Fig. 3.4 Effect of different pH on surface tension and emulsification activity

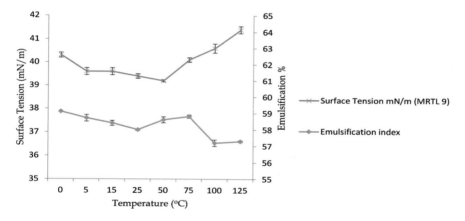

Fig. 3.5 Effect of different temperature on surface tension and emulsification activity

Biosurfactant obtained from *L. paracasei* evaluated for their stability at different pH and temperature ranges from 4 to 10 and 25, 37 and 60 °C, respectively. The stability of the biosurfactant produced at different pH and temperature was determined by observing the change in surface tension. About 50 mg/mL concentration biosurfactant sample was prepared and incubated at different pH (4.0–10.0) and temperature (25, 37 and 60 °C). The minimal value of surface tension was observed at pH 7.0, i.e., 41.8 mN/m. The surface tension of the biosurfactant persisted or somewhat stable to pH deviations between pH 6.0 and 10.0. It was found that biosurfactant is highly stable at alkaline than acidic environment (Gudiña et al. 2010). Subsequently, biosurfactant was found to be stable after treatment at 25, 37, and 60 °C for 120 h with no noteworthy loss of activity.

The synergistic effect of pH, salinity, and temperature on the surface active properties of the biosurfactant obtained from the *L. pentosus* (Vecino Bello et al. 2012) were evaluated. Box-Behenken factorial design was used as a tool to study the synergistic effect of pH, salinity, and temperature on biosurfactant production. It has been concluded that the factor affected most was pH, trailed by salinity and temperature showed insignificant effect within the range tested. Whereas, the factor that had maximal effects on emulsification potential was also pH, trailed by temperature and salinity. Lower salinity permits the use of biosurfactant at low pH. Salinity significantly declined the surface active properties of the biosurfactants and also limits their application in bioremediation where salinity is the vital component. Augustine et al. (2010) reported that biosurfactant obtained from *Lactobacillus* spp. was found to be stable at different temperatures tested in conviction of surface tension and interfacial tension. On the other hand, biosurfactant stability at different NaCl concentrations was also evaluated and found to be stable at different NaCl concentrations in terms of their emulsification activity.

Toxicity

In current scenario, there is a vast apprehension concerning the toxicity evaluation of biosurfactants used for biomedical and food applications. So, there is a prompt necessity to spot toxicity of biosurfactants. The mouse fibroblasts cells are generally used for cytotoxicity assessment for pharmaceutical and other therapeutic material. These cell lines originated from the mouse fibroblast cells are advised for in vitro assessment of toxicity for biomedical materials. Numerous reports on cytotoxic assessment of biosurfactant reported that the absence of cytotoxicity is anticipated while you wish for environmentally safe chemicals. Characteristically, the cytotoxicity appears associated to its interactions with the phospholipids of plasma membrane and consequently cell lysis (Sharma et al. 2015). Cytotoxicity of biosurfactants produced by *E. faecium* MRTL9 has been assayed using mouse fibroblast (ATCC L929) cell line (Sharma et al. 2015).

Throughout cytotoxicity evaluation, various concentrations of biosurfactant and commercially available rhamnolipids (Janeil, USA) were dissolved in DMSO. On the contrary, SDS at equivalent concentration acted as a negative control. Noteworthy variances in fibroblast cell viability were witnessed at different concentrations of biosurfactants ranging 25–6.25 mg ml^{-1}. Maximum 90 % cell viability of muse fibroblast cells was foundat 6.25 mg ml^{-1} with biosurfactant produced by *E. faecium* MRTL9, while commercial rhamnolipid showed 35.33 % viability approximately equivalent to SDS, i.e., 35 %. However, increase in the amount of biosurfactant dropped the cell line viability Table 3.3.

Table 3.3 Effect of biosurfactant derived from *E. faecium* MRTL9 on different growth parameters of *T. aestivum*

Biosurfactant concentration	Seed germination	Root elongation	Germination index	Vigor index
½ CMC	100 ± 0.2	112 ± 0.3	112 ± 0.23	1250 ± 125
CMC	100 ± 0.1	110 ± 0.2	110 ± 0.5	1435 ± 140
CMC	100 ± 0.15	117 ± 0.3	117 ± 0.12	1500 ± 120
Distilled water	100 ± 0.1	120 ± 0.15	120 ± 0.15	1495 ± 128
SDS	18 ± 0.2	22 ± 0.19	22 ± 0.5	600 ± 115

The present study confirmed that the biosurfactant had lower toxicity as compared to SDS and rhamnolipid (Fig. 3.6). SDS has been appreciated as a reference irritant for its nonallergenic and toxicity behavior (Effendy and Howard 1996).

Sambanthamoorthy et al. (2014) obtained biosurfactant from *Lactobacillus jensenii* and *Lactobacillus rhamnosus* and evaluated their toxicity against human lung epithelial cell line (A549). The cytotoxicities observed for the crude biosurfactants from *L. jensenii* and *L. rhamnosus* were evaluated on eukaryotic cells by the discharge of lactate dehydrogenase (LDH) and the overall cell number examine. During the cytotoxicity assessment of the biosurfactant at various concentrations incubated for 24 h, the cell-free supernatant was collected for LDH concentration determination. Biosurfactants derived from *L. jensenii* and *L. rhamnosus* were observed to have very low toxicity levels at 200 mg/mL. Biosurfactant concentrations ranging from 25 to 100 mg/mL showed no toxicity.

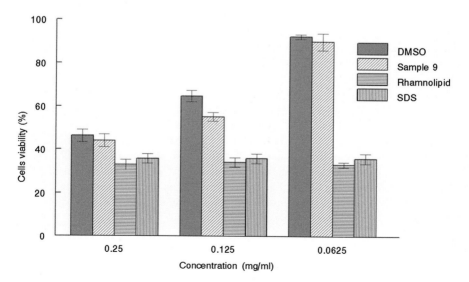

Fig. 3.6 Effect of biosurfactant on cell viability of mouse fibroblast cell in order to determine the cytotoxicity

Phytotoxicity

Phytotoxicity is also an important parameter to determine the toxicity evaluations for the biosurfactants produced by various microorganisms. The effect of biosurfactants on static seed germination and root elongation was observed by various researchers on different plants like *Brassica nigra* and *Triticum aestivum* (Sharma et al. 2015). Solutions of biosurfactant obtained from the *E. faecium* MRTL9 were prepared in distilled water at different concentrations ranging from 1.25 to 5 mg/mL. And it was observed that the biosurfactant derived from *E. faecium* MRTL9 is a nonphytotoxic compound. The germination test has been used in determining phytotoxicity due to its little implementation cost and ease of performance. The previous works of various authors report that some surfactants have an inhibitory effect on plant growth (Binks et al. 1991). Nearly 100 % seed germination was recorded in both types of seeds treated with the biosurfactant produced by *E. faecium*. Vital growth parameters such as root elongation, vigor index, and germination index were found superior in case of biosurfactant treatment as compared to SDS. Furthermore, all the vital growth parameters were found increasing with increase in the concentration of biosurfactant.

Conclusion

Microbial surfactants show various characteristics which could be beneficial in many formulations of food processing. Recently, their antibiofilm properties have engrossed consideration as an approach to inhibit or disturb the biofilms formed in food processing surfaces. The blend of properties like emulsifying, antibiofilm, and antimicrobial attributed by the biosurfactants recommends the potential application as versatile additives. Inadequate information concerning toxicity seems to be the major reason for the restricted uses of biosurfactants in food and therapeutics. Biosurfactants derived from GRAS microorganisms such as probiotic Lactobacilli are of abundant potential for food and biomedical applications but, much more research is requisite on this field. In view of novel surface active molecules derived from microorganisms can contribute for the different molecules in terms of structural composition, but the toxicological fate of novel biosurfactants should be underlined in order to endorse these compounds for food processing. Considering the community and technological advancement, utilization of biosurfactants, which are less toxic and extremely functional, has more significance and should be developed.

References

Augustine N, Kumar P, Thomas S (2010) Inhibition of Vibrio cholerae biofilm by AiiA enzyme produced from *Bacillus* sp. Arch Microbiol 192:1019–1022

Banat IM, Franzetti A, Gandolfi I, Bestetti G, Martinotti MG, Fracchia L, Marchant R (2010) Microbial biosurfactants production, applications and future potential. Appl Microbiol Biotechnol 87(2):427–444

Binks, BP, Fletcher PDI, Horsup DI (1991) Effect of microemulsified surfactant in destabilising water-in-oil emulsions containing C12E4. Colloids and surfaces 61:291–315

Busscher HJ, Neu TR, Van der Mei HC (1994) Biosurfactant production by thermophilic dairy streptococci. Appl Microbiol Biotechnol 41(1):4–7

Bustos G, De la Torre N, Moldes AB, Cruz JM, Domínguez JM (2007) Revalorization of hemicellulosic trimming vine shoots hydrolyzates trough continuous production of lactic acid and biosurfactants by L. pentosus. J Food Eng 78(2):405–412

Cirigliano MC, Carman GM (1985) Purification and characterization of liposan, a bioemulsifier from *Candida lipolytica*. Appl Environ Microb 50(4):846–850

Desai JD, Banat IM (1997) Microbial production of surfactants and their commercial potential. Microbiol Mol Biol Rev 61(1):47–64

Effendy I, Howard IM (1996) Detergent and skin irritation. Clin Dermatol 14(1):15–21

Fracchia L, et al (2010) A *Lactobacillus*-derived biosurfactant inhibits biofilm formation of human pathogenic *Candida albicans* biofilm producers. Appl Microbiol Biotechnol 2:827–837

George S, Jayachandran K (2009) Analysis of rhamnolipid biosurfactants produced through submerged fermentation using orange fruit peelings as sole carbon source. Appl Biochem Biotechnol 158(3):694–705

Goek P, Bednarski W, Brzozowski B, Dziuba B (2009) The obtaining and properties of biosurfactants synthesized by bacteria of the genus *Lactobacillus*. Ann Microbiol 59(1):119–126

Gudiña EJ, Teixeira JA, Rodrigues LR (2010) Isolation and functional characterization of a biosurfactant produced by *Lactobacillus paracasei*. Colloids Surf B 76(1):298–304

Gudiña EJ et al (2013) Potential therapeutic applications of biosurfactants. Trends Pharmacol Sci 34(12):667–675

Kachholz T.R.A.U.D.E.L, Schlingmann M (1987) Possible food and agricultural applications of microbial surfactants: an assessment. Biosurfactants and biotechnology 183–210

Lima, Tânia MS, et al (2011) Evaluation of bacterial surfactant toxicity towards petroleum degrading microorganisms. Bioresour Technol 102(3):2957–2964

Liu X, Ren B, Chen M, Wang H, Kokare CR, Zhou X, Zhang L (2010) Production and characterization of a group of bioemulsifiers from the marine *Bacillus velezensis* strain H3. Appl Microbiol Biotechnol 87(5):1881–1893

Madhu AN, Prapulla SG (2014) Evaluation and functional characterization of a biosurfactant produced by *Lactobacillus plantarum* CFR 2194. Appl Biochem Biotechnol 172(4):1777–1789

Moldes AB, Paradelo R, Vecino X, Cruz JM, Gudiña E, Rodrigues L, Barral MT (2013) Partial characterization of biosurfactant from *Lactobacillus pentosus* and comparison with sodium dodecyl sulphate for the bioremediation of hydrocarbon contaminated soil. BioMed Res Int 961842:6 p

Mulligan CN (2005) Environmental applications for biosurfactants. Environ Pollut 133(2):183–198

Mulligan CN, Yong RN, Gibbs BF (2001) Surfactant-enhanced remediation of contaminated soil: a review. Eng Geol 60(1):371–380

Muthusamy K, et al (2008) Biosurfactants: properties, commercial production and application. Current Science 94(6):00113891

Myers D (2005) The organic chemistry of surfactants. Surfactant Science and Technology, 3rd edn. Wiley, Hoboken, pp 29–79. doi:10.1002/047174607X.ch2

Neu TR, Karl P (1990) Emulsifying agents from bacteria isolated during screening for cells with hydrophobic surfaces. Appl Microbiol Biotechnol 32(5):521–525

Nitschke M, Pastore GN (2006) Production and properties of a surfactant obtained from *Bacillus subtilis* grown on cassava wastewater. Bioresour Technol 97(2):336–341

Nitschke M, Costa SGVAO (2007) Biosurfactants in food industry. Trends Food Sci Technol 18 (5):252–259

Portilla-Rivera O, Torrado A, Domínguez JM, Moldes AB (2008) Stability and emulsifying capacity of biosurfactants obtained from lignocellulosic sources using *Lactobacillus pentosus*. J Agric Food Chem 56(17):8074–8080

Rivera OMP, Moldes AB, Torrado AM, Domínguez JM (2007) Lactic acid and biosurfactants production from hydrolyzed distilled grape marc. Process Biochem 42(6):1010–1020

Rodríguez N, et al (2010) Alternatives for biosurfactants and bacteriocins extraction from *Lactococcus lactis* cultures produced under different pH conditions. Lett Appl Microbiol 51 (2):226–233

Rodrigues L, Van der Mei HC, Teixeira J, Oliveira R (2004) Influence of biosurfactants from probiotic bacteria on formation of biofilms on voice prostheses. Appl Environ Microbiol 70(7):4408–4410

Rodrigues L, Banat IM, Teixeira J, Oliveira R (2006) Biosurfactants: potential applications in medicine. J Antimicrob Chemother 57(4):609–618

Rodríguez-Pazo N, Salgado JM, Cortés-Diéguez S, Domínguez JM (2013) Biotechnological production of phenyllactic acid and biosurfactants from trimming vine shoot hydrolyzates by microbial co-culture fermentation. Appl Biochem Biotechnol 169(7):2175–2188

Saharan BS, Sahu RK, Sharma D (2011) A review on biosurfactants: fermentation, current developments and perspectives. Genetic Eng Biotechnol J 1:1–14

Sambanthamoorthy K, et al (2014) Antimicrobial and antibiofilm potential of biosurfactants isolated from lactobacilli against multi-drug-resistant pathogens. BMC Microb 14(1):1

Sharma, Deepansh, et al (2014) Production and structural characterization of *Lactobacillus helveticus* derived biosurfactant. The Scientific World J 2014

Sharma, Deepansh, et al (2015) Isolation and functional characterization of novel biosurfactant produced by *Enterococcus faecium*. SpringerPlus 4(1):1–14

Thavasi R, Jayalakshmi S, Banat IM (2011) Effect of biosurfactant and fertilizer on biodegradation of crude oil by marine isolates of *Bacillus megaterium*, *Corynebacterium kutscheri* and *Pseudomonas aeruginosa*. Bioresour Technol 102(2):772–778

Van der Vegt W, Van der Mei HC, Noordmans J, Busscher HJ (1991) Assessment of bacterial biosurfactant production through axisymmetric drop shape analysis by profile. Appl Microbiol Biotechnol 35(6):766–770

Vecino Bello X, Devesa-Rey R, Cruz JM, Moldes AB (2012) Study of the synergistic effects of salinity, pH, and temperature on the surface-active properties of biosurfactants produced by *Lactobacillus pentosus*. J Agric Food Chem 60(5):1258–1265

Vecino X, et al (2013) Evaluation of biosurfactant obtained from *Lactobacillus pentosus* as foaming agent in froth flotation. J Environ Manage 128:655–660

Vecino X, Barbosa-Pereira L, Devesa-Rey R, Cruz JM, Moldes AB (2014) Study of the surfactant properties of aqueous stream from the corn milling industry. J Agric Food Chem 62(24): 5451–5457

Velraeds MM, Van der Mei HC, Reid G, Busscher HJ (1996) Inhibition of initial adhesion of uropathogenic *Enterococcus faecalis* by biosurfactants from *Lactobacillus* isolates. Appl Environ Microbiol 62(6):1958–1963

Chapter 4
Structural Properties of Biosurfactants of Lab

Abstract Unlike surfactants of chemical origin, which are characterized according to their nature, biosurfactants are generally characterized mostly by their chemical composition and origin. Biosurfactants are composed of hydrophilic moiety (polar, head) containing of amino acids, peptides, polysaccharides; and hydrophobic moiety (nonpolar tail) composed of fatty acids. The main cause that confines its commercialization is the inadequate knowledge of structural composition, so as to limiting its application as pharmaceuticals and therapeutic agents. Generally, biosurfactants obtained from LAB are found as multicomponent mixtures composed of polysaccharides, lipids, phosphate groups, and proteins. Structural properties are vital to design customized biosurfactants based on specific applications. Still, only inadequate information is available in the literature related to the biosurfactants derived from the LAB. Development in advanced chromatographic techniques, enhanced purification of the biosurfactant, and production from simpler medium could lead to determine structurally purified biosurfactants for future applications in food and therapeutics formulations.

Keywords Biosurfactant · Glycolipids · Glycoproteins · Xylolipids · Fatty acids

Introduction

Unlike surfactants of chemical origin, which are characterized according to their nature, biosurfactants are generally characterized mostly by their chemical composition and origin. Biosurfactants composed of hydrophilic moiety (polar, head) consist of amino acids, peptides, polysaccharides; and hydrophobic moiety (non-polar tail) composed of fatty acids. Hence, the major classification of biosurfactant derived from the microbial cell includes glycolipids, lipopeptides, phospholipids, fatty acids, and particulate surfactants (Fig. 4.1). The most promising group of biosurfactants are the glycolipids, this group will be discussed more in detail.

© The Author(s) 2016
D. Sharma et al., *Biosurfactants of Lactic Acid Bacteria*,
SpringerBriefs in Microbiology, DOI 10.1007/978-3-319-26215-4_4

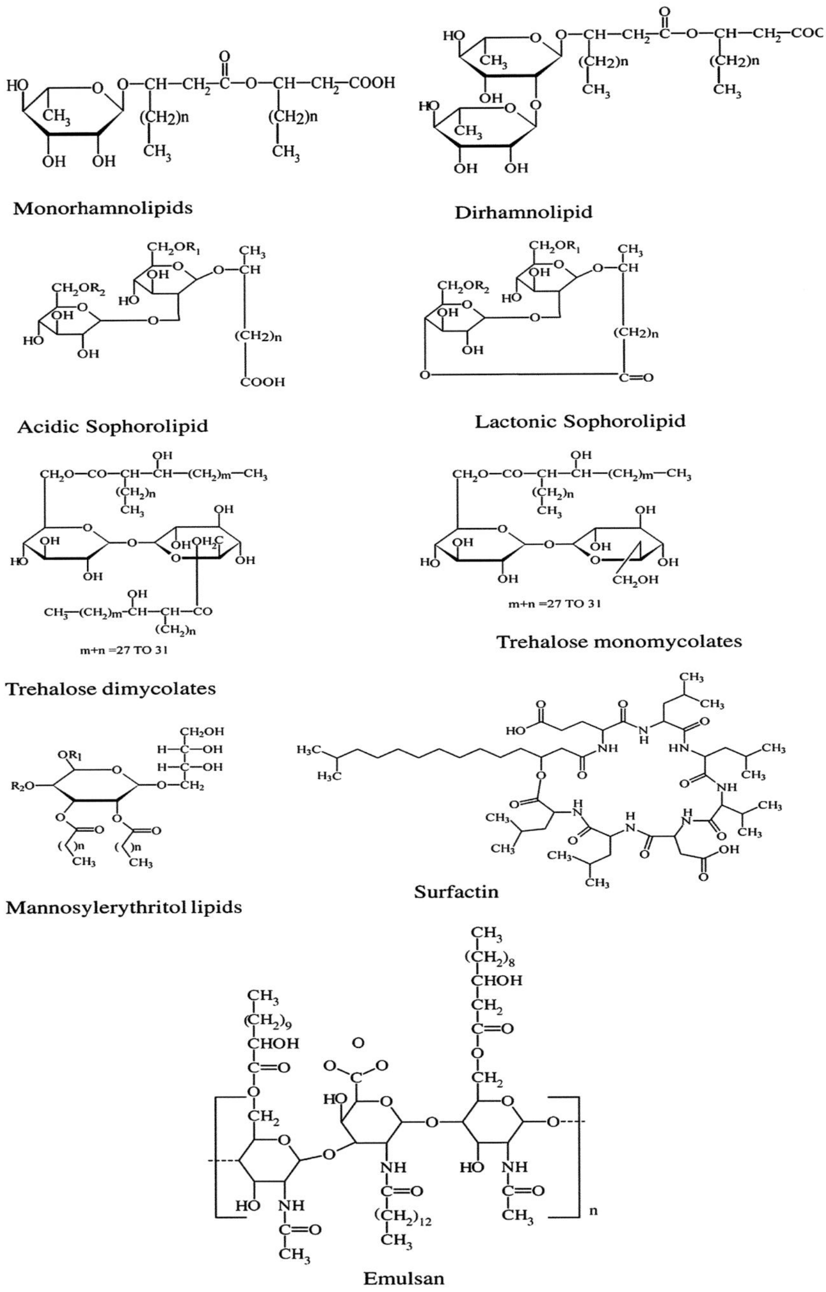
Monorhamnolipids
Dirhamnolipid
Acidic Sophorolipid
Lactonic Sophorolipid
m+n =27 TO 31
Trehalose dimycolates
m+n =27 TO 31
Trehalose monomycolates
Mannosylerythritol lipids
Surfactin
Emulsan

Fig. 4.1 Classification and structure of various biosurfactants reported (*Courtesy* Perfumo et al. 2010)

Glycolipids

Glycolipids generally occur in combination of carbohydrates with certain fatty acids, hydroxy fatty acids. As of their high yield and the potential to utilize low-cost substrates for commercial production, there is increased interest for glyolipids in scientific community (Saharan et al. 2011). The most considered glycolipids molecules are rhamnolipids obtained from *Pseudomonas* spp., mannosylerythriol biosurfactants obtained from *Pseudozyma antarctica* (Kitamoto et al. 1990), trehalose lipids derived from *Rhodococcus* sp., *Nocardia* sp., and *Mycobacterium* sp. and sophorolipids produced from *Candida* spp. (Cooper and Paddock 1984).

Rhamnolipids

Rhamnolipids are the kind of glycolipids in which rhamnose sugars are linked to one or two molecules of β-hydroxydecanoic fatty acid. Production of rhamnose-containing glycolipids was described in *Pseudomonas aeruginosa* (Toribio et al. 2011; Nitschke et al. 2005). Rhamnolipids derived from *P. aeruginosa* is the most considered producer with more than 100 g/L production can be attained. This comparatively high production amount makes rhamnolipids most striking kind of biosurfactants for added commercial value. Similarly, sophorolipids, rhamnolipids are mixture of analogous molecules (Syldatk et al. 1985). Rhamnolipids decrease the surface tension of water from 72.80 to 27 mN/m. As of their significant biocompatibility and biodegradability, rhamnolipids can be used in various commercial and environmental applications such as petroleum recovery (Parra et al. 1989) control of marine oil spills and soil washing or remediation (Banat 1995). The rhamnolipid biosurfactants are assumed to interact with the phosphatidylethanolamine of biological plasma membrane (Sanchez et al. 2006). A matter of apprehension in the commercialization of rhamnolipids is the fact that *P. aeruginosa* is an opportunistic pathogen. So there is immense need of the hour that microorganism with GRAS status could be used for industrial production of the biosurfactant for their better prospecting in food and therapeutics application.

Trehalolipids

Various microorganisms produce glycolipids with trehalose sugar moieties (α-1,1 glucose disaccharide) like *Mycobacterium* sp., *Corynebacterium* sp., and *Rhodococcus* spp. (Desai and Banat 1997). Trehalolipids from various microorganisms diverge in the size and structure of mycolic acid, the number of carbon atoms presents and the amount of unsaturation (Asselineau and Asselineau 1978). Trehalose dimycolate obtained from *Rhodococcus erythropolis* has been extensively studied (Rapp et al. 1979).

Sophorolipids

Sophorolipids, which are obtained from yeasts such as *Torulopsis bombicola* (Cooper and Paddock 1984) are composed of a carbohydrate (sophorose) linked to hydroxy fatty acid chains. Analogous mixtures of water-soluble sophorolipids obtained from various yeast strains have been documented (Hommel et al. 1987). *T. petrophilum* derived sophorolipids on water-insoluble substrates like alkanes and vegetable oils. Sophorolipids, which were chemically alike to those derived from *T. bombicola*, did not emulsify vegetable oils. Sophorolipids reduced the interfacial tension between *n*-hexadecane and water and exhibited significant stability toward pH and temperature changes (Cooper and Paddock 1983).

Lipopeptides and Lipoproteins

The molecules in lipopeptide and lipoprotein biosurfactants contain cyclic peptides linked to long chains of fatty acid. Various microorganisms are recognized to produce lipopeptides and lipoproteins, e.g., *Bacillus subtilis*. Biosurfactant derived from the *B. subtilis* are very potent which reduced the surface tension from 72.8 to 27.9 mN/m at very low concentrations of 0.005 % (Arima et al. 1968). Additionally, the biosurfactant derived from *B. subtilis* possesses antibacterial, antiviral, antiadhesive, antimycoplasma, and hemolytic properties. An aminolipid biosurfactant named serratamolide has been obtained from *Serratia marcescens* (Matsuyama et al. 1986). Lipopeptides and lipoproteins increased cell hydrophilicity by barricading the hydrophobic moiety on cell surfaces.

Fatty Acids and Phospholipids

Several microorganisms are capable to grow on various hydrophobic substrates, discharge huge quantities of phospholipids, fatty acids or neutral lipids to permit the uptake of the carbon source (Käppeli and Finnerty 1979; Robert et al. 1989). Phosphatidylethanolamine derived from *R. erythropolis* grown on *n*-alkane reduced the interfacial tension among water and hexadecane to less than 1 mN/m with a CMC of 30 mgL^{-1} (Kretschmer et al. 1982).

Polymeric Biosurfactants

Emulsan, derived from *Acinetobacter calcoaceticus*, is the best observed and studied biosurfactant. Emulsan mainly composed of a heteropolysaccharide backbone to which fatty acids are covalently linked. A different illustration is liposan, which is composed of carbohydrate–protein complex derived from the yeast *Yarrowia lipolytica* (Cirigliano and Carman 1984). *A. calcoaceticus* RAG-1 excreted amphipathic heteropolysaccharide bioemulsifier known as emulsan. The fatty acids present are covalently linked to the polysaccharide moiety through *o*-ester linkages (Shabtai and Gutnick 1985). Likewise, an extracellular biosurfactant composed of mixture of carbohydrate, protein, and lipids was found from *C. tropicalis* (Singh et al. 1990)

Structural Properties of Biosurfactant of LAB

The main cause that confines its commercialization is the inadequate knowledge of structural composition, so as to its restrictive application as pharmaceuticals and therapeutic agents. Several studies on biosurfactants obtained from lactobacilli have been executed, but less is identified about their structure, possibly due to high complexity. Biosurfactants derived from various LAB have been categorized as complex mixtures that constrain the adhesion of pathogenic microbes to biotic and abiotic surfaces, but their chemical composition has not been comprehensively reported and only a few have been incompletely described (Sharma et al. 2015a, b; Velraeds et al. 1996; Rodrigues et al. 2006). Improved information of biosurfactant's composition obtained from LAB is vital to effusively recognize their active components and be able to transform them in order to enhance their properties (Table 4.1).

Biosurfactants derived from various LAB were identified as multicomponent compounds, containing of protein and polysaccharides, perhaps comprising of phosphate groups (Velraeds et al. 1996; Rodrigues et al. 2006). The biosurfactant derived from *S. thermophilus* B was described as a mixture of different constituents, comprising polysaccharides and glycolipids (Busscher et al. 1997). Biosurfactant derived from *S. thermophilus* B was found as an effective surface active molecule, reducing the surface tension of water to 35 mN/m. Similar structural properties have also been reported with biosurfactants produced by two different *Streptococcus mitis* which are composed of glycolipids with low level of proteins (Van Hoogmoed et al. 2004). Biosurfactants secreted by two different *S. mitis* recognized as a rhamnolipid-like molecule decreased the surface tension to 35 mN/m at a concentration of 1 mg/ml. Gudiña et al. (2010) isolated biosurfactant derived from *L. paracasei* has been characterized multicomponent biosurfactant.

It was found that biosurfactant produced by *L. lactis* is composed of protein and polysaccharide and phosphate group. Various collagen-binding proteins in the

Table 4.1 Structural composition of biosurfactants derived from various LAB strains

S. no.	Strain	Biosurfactant produced	References
1.	*L. acidophilus* RC14	Rich in protein, high amount of polysaccharides and Phosphate content	Velraeds et al. (1996)
2.	*S. thermophilus*	Glycolipid	Busscher et al. (1997)
3.	*L. acidophilus*	Surlactin	Velraeds et al. (1996)
4.	*S. mutans* NS	Rhamnolipid like	Van Hoogmoed et al. (2004)
5.	*S. thermophiles* A	Glycolipid	Rodrigues et al. (2006)
6.	*L. casei*	Glycoprotein	Golek et al. (2009)
7.	*L. lactis*	Xylolipids	Saravanakumari and Mani (2010)
8.	*L. acidophilus*	Glycoprotein	Tahmourespour et al. (2011)
9.	*L. plantarum*	Glycolipids	Sauvageau et al. (2012)
10.	*L. plantarum*	Glycoprotein	Madhu and Prapulla (2014)
11.	*L. pentosus*	Glycolipids	Vecino et al. (2014)
12.	*L. casei* MRTL3	Glycolipids	Sharma and Saharan (2014)
13.	*E. faecium* MRTL9	Xylolipids	Sharma et al. (2015a, b)
14.	*L. helveticus* MRTL91	Glycolipids (Xylolipids)	Sharma et al. (2015a, b)
15.	*L. pentosus*	Glycolipopeptide	Vecino et al. (2015)

biosurfactants derived from numerous LAB strains, biosurfactant derived from *L. fermentum* RC-14 was found to contain proteins responsible for the antibiofilm activity against *Enterococcus faecalis*.

Purified biosurfactant isolated from *Lactococcus lactis* appeared as crystalline white mixture and found as anionic in nature (Saravanakumari and Mani 2010). Fatty acids composition determination using Gas chromatography-Mass spectroscopy (GC-MS) of the biosurfactant obtained from *L. lactis* confirmed the occurrence of octadecanoic acid as a major fatty acid present in the sample. GC-MS spectrum analysis of biosurfactant obtained from *L. lactis* confirmed peaks for xylopyranoside sugar and octadecanoic acid as a fatty acid. Xylosepyranoside octadecanoic acid containing biosurfactants is related to glycolipids present in

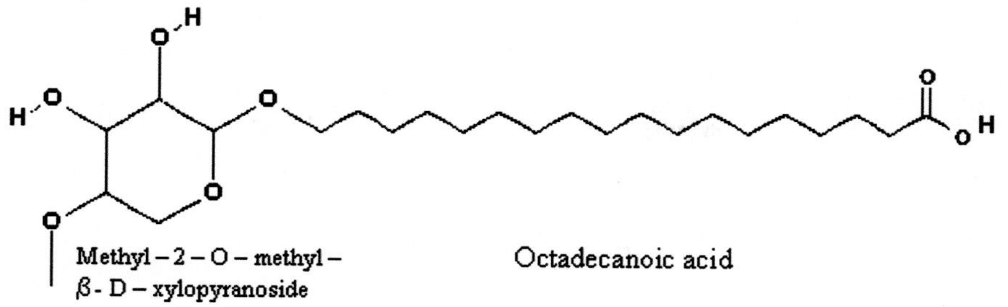

Methyl – 2 – O – methyl –   Octadecanoic acid
β- D – xylopyranoside

Fig. 4.2 Xylolipid reported from the *L. lactis* (*Courtesy by* Saravanakumari and Mani 2010)

rhamnolipid derived from *P. aeruginosa*. GC-MS analysis and ^{1}H NMR ratify that the biosurfactant obtained from probiotic bacteria is a glycolipid biosurfactant (Fig. 4.2) and known as xylolipid.

Sauvageau et al. (2012) isolated glycolipids from *L. plantarum* IRL 560 and characterized their structural composition as α-$_D$-Glcp-diglyceride (GL1), α-$_D$-Galp-(1-2)- α-$_D$-Glcp-diglyceride (GL2a), β-D-Glcp-(1-6)- α-$_D$-Galp-(1-2)-6-O-acyl- α-$_D$-Glcp-diglyceride (GL2b), and β-DGlcp-(1-6)-a-D-Galp-(1-2)- α-$_D$-Glcp-diglyceride (GL3) (Fig. 4.3). These glycolipid arrangements were also found in *L. casei*. With the understanding of the definite confirmation for the arrangements of the produced glycolipids, this will aid in relating the biological properties of *L. plantarum*.

NMR and FTIR Analysis of LAB Derived Biosurfactants

FTIR analysis recognizes chemical constituents in any chemical formulations such as paints, polymers, biosurfactants, bioemulsifiers, pharmaceuticals, and food-grade materials. FTIR has been frequently used as a vital tool to determine the position of carbon and hydrogen in glycolipids. Various studies had demonstrated the applicability of these tools to determine the biosurfactant structural attributes. Biosurfactant derived from the LAB were described as multicomponent mixtures containing protein fractions, various polysaccharides and phosphate groups on the basis of FTIR analysis (Rodrigues et al. 2004; Thavasi et al. 2008; Falagas and Makris 2009; Sauvageau et al. 2012; Madhu and Prapulla 2014). Biosurfactant obtained from the *E. faecium* MRTL 9 was structurally characterized as glycolipid (Fig. 4.4, Table 4.2). Exploration of FTIR spectrum of biosurfactant showed the structure as lipid amalgamated with polysaccharide fractions. The functional group determination by FTIR exposed that most prominent adsorption bands were positioned at 3009, 2855, and 2955 cm^{-1} (C-H stretching of CH$_2$ and CH$_3$ groups), whereas absorption bands at 1674 cm^{-1} (C = O stretching of the carbonyl groups), 1103 cm^{-1} (C-O stretching bands between carbon and hydroxyl groups), and 771 cm^{-1} (CH$_2$ group) which obviously established the occurrence of glycolipid type of biosurfactant

GL1

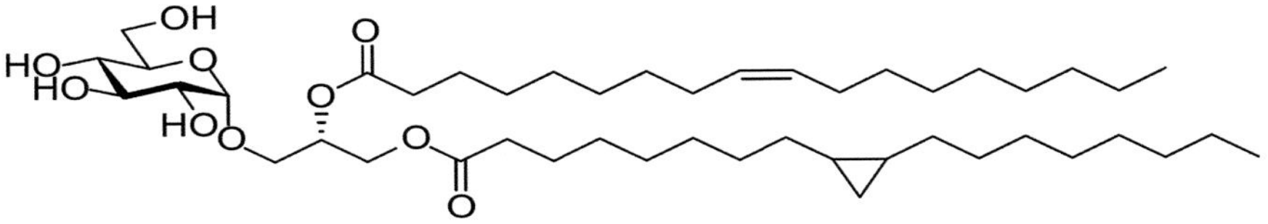

GL2a

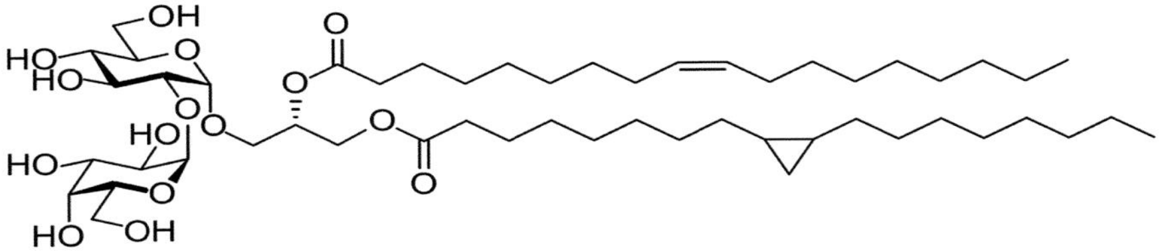

GL2b

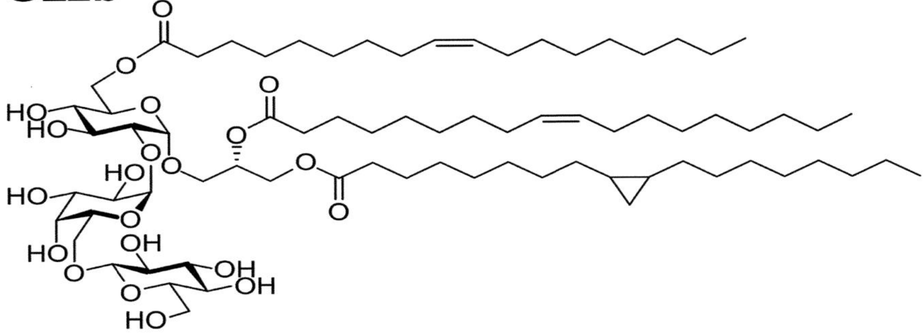

GL3

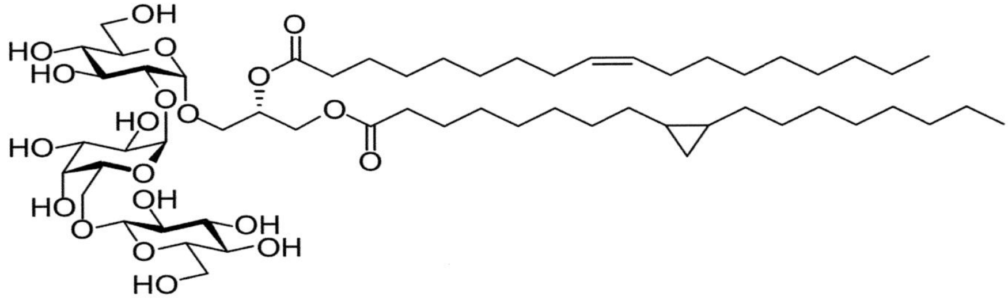

Fig. 4.3 Sauvageau et al. (2012) isolated glycolipids from *L. plantarum* IRL 560

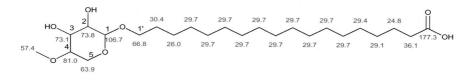

Fig. 4.4 Xylolipids from *E. faecium* MRTL9 (Sharma et al. 2015a, b)

Table 4.2 Different absorption peaks for xylolipids from *E. faecium* MRTL9	Absorbance range (cm^{-1})	Functional groups
	3000–3600	OH
	2900–2950	C-H groups CH$_2$ and CH$_3$
	1400–1460	C = H
	1725, 1675	C = O
	1000–1300	C-O
	675–1000	C-H

(Thavasi et al. 2008). The hydrophobic tail of biosurfactant is mainly composed of long chain of fatty acids. Similar data has been reported for other lactobacilli like *L. lactis* producing the xylolipids (glycolipid) (Saravanakumari and Mani 2010).

NMR analysis also determines the hydrogen and carbon atom spatial arrangement in various biological origin products like biosurfactants. In few cases of biosurfactant obtained from the LAB, NMR has been used as analytical tool for determination of their structural attributes. The spatial organization of proton and carbon in biosurfactant has been carried out by ^{1}H NMR and ^{13}C NMR spectra (Saravanakumari and Mani 2010; Sharma and Saharan 2014; Sharma et al. 2014; Saharan et al. 2014). The distinctive chemical shifts pointed out that the biosurfactant has the structural arrangement closely similar to glycolipid (Table 4.3). Various specialized peaks were reported in NMR spectra of biosurfactants derived from LAB representing presence of polysaccharides and long aliphatic fatty acids chains. The incidence of chemical shift (δ) at 4.09, 3.53, 3.53, 2.35, 3.72, and 3.52 ppm designated C1, C2, C3, C4, C5, OCH3, and -COOH groups, respectively. Specific spectra peaks of NMR were also described in xylolipid obtained from

Table 4.3 NMR spectra of xylolipids from *E. faecium* MRTL9	Assignment (s)	1H NMR (ppm)	13C NMR (ppm)
	C-1	4.09	127.38
	C-2	3.53	∼60
	C-3	3.53	∼60
	C-4	2.35	69.5
	C-5	3.72, 3.61	58.81
	OCH$_3$	3.52	53.51
	COOH	–	170.79

Lactococcus lactis (Saravanakumari and Mani 2010; Morita et al. 2012; Partovi et al. 2013; Desai and Banat 1997; Falagas and Makris 2009).

Fatty Acids

Biosurfactants are made up of hydrophobic and hydrophilic moiety. Hydrophilic moiety is made up of the carbohydrate and hydrophobic moiety generally composed of the long chain of fatty acids. Several fatty acids have been reported as a part of hydrophobic moiety in biosurfactants derived from the LAB (Table 4.4). According to Sharma et al. (2015a, b), the fatty acid content of biosurfactant was determined by GC-MS. It was found that the biosurfactant obtained from the *E. faecium* MRTL9 mainly comprised of long-chain fatty acids. The main fatty acid of biosurfactants derived was a C-16 fatty acid, i.e., hexadecanoic fatty acid (Fig. 4.5). Saravanakumari and Mani (2010) also reported biosurfactant composed of octadecanoic acid as a major fatty acid chain linked with polysaccharide moiety from *L. lactis*.

The fatty acids derived from the *L. pentosus* were first converted into fatty acid methyl esters (FAMEs) and then observed using GC-MS. Fatty acids analysis revealed that the main fatty acids confined were linoelaidic acid trailed by oleic acid or elaidic acid. And it was found motivating that the fatty acid chain linked with the biosurfactant was quiet similar to the fatty acid chains existing in rhamnolipid derived from the *P. putida* (Vecino et al. 2015). The most studied glycolipid i.e. Rhamnolipids are contains β-hydroxydecanoic acid as a long fatty acid chain (Desai and Banat 1997). In a recent study, the glycolipids of *L. plantarum* revealed that it contains various fatty acids in different kind of glycolipids found in biosurfactant. Palmitic acid, oleic acid and octadecanoic acid are the major fatty acids present in the biosurfactant produced by the *L. plantarum* (Sauvageau et al. 2012).

Table 4.4 Fatty acid composition of biosurfactants isolated from various LAB strains

S. no.	Strain	Fatty acid	References
1.	*Lactococcus lactis*	Octadecanoic acid	Saravanakumari and Mani (2010)
2.	*E. faecium*	Hexadecanoic acid	Sharma et al. (2015a, b)
3.	*L. helveticus*	Hexadecanoic acid	Sharma et al. (2014)
4.	*L. pentosus*	Linoelaidic acid, oleic acid, palmitic acid and stearic acid	Vecino et al. (2015)
5.	*L. fermentum*	Hexadecanoic acid	Unpublished work
6.	*L. plantarum*	Palmitic acid, oleic acid and octadecanoic acid	Sauvageau et al. (2012)

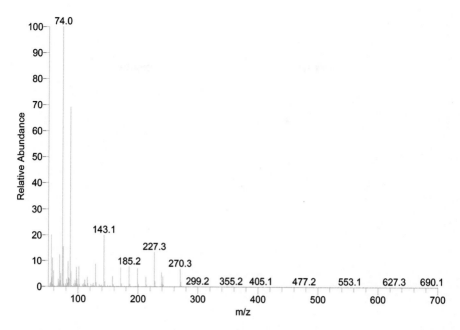

Fig. 4.5 Hexadecanoic acid as a main fatty acid of xylolipids produced by the *E. faecium* MRTL9 (Sharma et al. 2015a, b)

Conclusion

In this present chapter, the biosurfactant derived from different LAB strains were discussed. The surface tension, emulsification activity, and the critical micelle concentration are the critical parameters to characterize a novel biosurfactant from different lactobacilli. Structural properties of biosurfactants obtained from different LAB have been characterized as the multicomponent mixtures mainly glycoproteins and glycolipids. The advancement of NMR, FTIR, and tools like GC-MS increased the better knowledge of structural arrangements within the biomolecules. Characterization of structural attributes enable researchers for further advancement of the biosurfactants research for tailor-made applications and utilization for therapeutic applications. Still, only inadequate information about the structure of biosurfactant is available. Further development and studies are the demand of near future to decode the structural characters of the biosurfactants.

References

Arima K, Kakinuma A, Tamura G (1968) Surfactin, a crystalline peptidelipid surfactant produced by *Bacillussubtilis*: Isolation, characterization and its inhibition of fibrin clot formation. Biochem Bioph Res Co 31(3):488–494

Asselineau C, Asselineau J (1978) Trehalose-containing glycolipids. Prog Chem Fats Lipids 16:59–99

Banat IM (1995) Biosurfactants production and possible uses in microbial enhanced oil recovery and oil pollution remediation: a review. Bioresource Technol 51(1):1–12

Busscher HJ, Van der Mei HC (1997) Physico-chemical interactions in initial microbial adhesion and relevance for biofilm formation. Adv Dent Res 11(1):24–32

Cirigliano MC, Carman GM (1984) Isolation of a bioemulsifier from *Candida lipolytica*. Appl Environ Microb 48(4):747–750

Cooper DG, Goldenberg BG (1983) Surface-active agents from two Bacillus species. Appl Environ Microbiol 53(2):224–229

Cooper DG, Paddock DA (1983) Torulopsis *petrophilum* and surface activity. Appl Environ Microbiol 46(6):1426–1429

Cooper DG, Paddock DA (1984) Production of a biosurfactant from Torulopsis bombicola. Appl Environ Microbiol 47(1):173–176

Desai JD, Banat IM (1997) Microbial production of surfactants and their commercial potential. Microbiol Mol Biol Rev 61(1):47–64

Falagas ME, Makris GC (2009) Probiotic bacteria and biosurfactants for nosocomial infection control: a hypothesis. J Hosp Infec 71(4):301–306

Gołek P, Bednarski W, Brzozowski B, Dziuba B (2009) The obtaining and properties of biosurfactants synthesized by bacteria of the genus *Lactobacillus*. Ann Microbiol 59(1): 119–126

Gudiña EJ, Teixeira JA, Rodrigues LR (2010) Isolation and functional characterization of a biosurfactant produced by *Lactobacillus paracasei*. Colloids Surf B 76(1):298–304

Hommel R, Stiiwer O, Stuber W, Haferburg D, Kleber HP (1987) Production of water-soluble surface-active exolipids by *Torulopsis apicola*. Appl Microbiol Biote 26(3):199–205

Käppeli O, Finnerty WR (1979) Partition of alkane by an extracellular vesicle derived from hexadecane-grown Acinetobacter. J Bacteriol 140(2):707–712

Kitamoto D, Yanagishita H, Shinbo T, Nakane T, Kamisawa C, Nakahara T (1993) Surface active properties and antimicrobial activities of mannosylerythritol lipids as biosurfactants produced by Candida antarctica. J Biotechnol 29(1):91–96

Kitamoto D, Akiba S, Hioki C, Tabuchi T (1990) Extracellular accumulation of mannosylery-thritol lipids by a strain of *Candida antarctica*. Agric Biol Chem 54(1):31–36

Kretschmer A, Bock H, Wagner F (1982) Chemical and physical characterization of interfacial-active lipids from *Rhodococcus erythropolis* grown on n-alkanes. Appl Environ Microb 44(4):864–870

Kuyukina MS, Ivshina IB, Philp JC, Christofi N, Dunbar SA, Ritchkova MI (2001) Recovery of Rhodococcus biosurfactants using methyl tertiary-butyl ether extraction. J Microbiol Methods 46(2):149–156

Madhu AN, Prapulla SG (2014) Evaluation and functional characterization of a biosurfactant produced by *Lactobacillus* plantarum CFR 2194. Appl Biochem Biotechnol 172(4):1777–1789

Maneerat S, Bamba T, Harada K, Kobayashi A, Yamada H, Kawai F (2006) A novel crude oil emulsifier excreted in the culture supernatant of a marine bacterium, Myroides sp. strain SM1. Appl Microbiol Biotechnol 70(2):254–259

Matsuyama T, Murakami T, Fujita M, Fujita S, Yano I (1986) Extracellular vesicle formation and biosurfactant production by *Serratia marcescens*. Microbiol 132(4):865–875

Morita T, Fukuoka T, Imura T, Kitamoto D (2012) Formation of the two novel glycolipid biosurfactants, mannosylribitol lipid and mannosylarabitol lipid, by *Pseudozyma parantarctica* JCM 11752T. Appl Microbiol Biotechnol 96(4):931–938

Morita T, Konishi M, Fukuoka T, Imura T, Kitamoto D (2006) Discovery of Pseudozyma rugulosa NBRC 10877 as a novel producer of the glycolipid biosurfactants, mannosylerythritol lipids, based on rDNA sequence. Appl Microbiol Biotechnol 73(2):305–313

Müller MM, Kügler JH, Henkel M, Gerlitzki M, Hörmann B, Pöhnlein M, ... Hausmann R (2012) Rhamnolipids—next generation surfactants? J Biotechnol 162(4):366–380

Nitschke M, Costa SG, Contiero J (2005) Rhamnolipid surfactants: an update on the general aspects of these remarkable biomolecules. Biotechnol Prog 21(6):1593–1600

Parra JL, Guinea J, Manresa MA, Robert M, Mercade ME, Comelles F, Bosch MP (1989) Chemical characterization and physicochemical behavior of biosurfactants. J Am Oil Chem Soc 66(1):141–145

Partovi M, Lotfabad TB, Roostaazad R, Bahmaei M, Tayyebi S (2013) Management of soybean oil refinery wastes through recycling them for producing biosurfactant using *Pseudomonas aeruginosa* MR01. World J Microbiol Biotechnol 29(6):1039–1047

Perfumo A, Smyth TJP, Marchant R, Banat IM (2010) Production and roles of biosurfactants and bioemulsifiers in accessing hydrophobic substrates. In Handbook of hydrocarbon and lipid microbiology, Springer, Berlin Heidelberg, pp 1501–1512

Rapp P, Bock H, Wray V, Wagner F (1979) Formation, isolation and characterization of trehalose dimycolates from *Rhodococcus erythropolis* grown on n-alkanes. Microbiol 115(2):491–503

Robert M, Mercade ME, Bosch MP, Parra JL, Espuny MJ, Manresa MA, Guinea J (1989) Effect of the carbon source on biosurfactant production byPseudomonas aeruginosa 44T1. Biotechnol Lett 11(12):871–874

Rodrigues L, Banat IM, Teixeira J, Oliveira R (2006) Biosurfactants: potential applications in medicine. J Antimicrob Chemother 57(4):609–618

Rodrigues L, Van der Mei H, Teixeira JA, Oliveira R (2004) Biosurfactant from *Lactococcus lactis* 53 inhibits microbial adhesion on silicone rubber. Appl Microbiol Biotechnol 66 (3):306–311

Rosenberg E, Ron EZ (1999) High-and low-molecular-mass microbial surfactants. Appl Microbiol Biotechnol 52(2):154–162

Saharan BS, Grewal A, Kumar P (2014) Biotechnological production of polyhydroxyalkanoates: a review on trends and latest developments. Chinese J Biol 2014

Saharan BS, Sahu RK, Sharma D (2011) A review on biosurfactants: fermentation, current developments and perspectives. Genetic Eng Biotechnol J 2011(1):1–14

Saravanakumari P, Mani K (2010) Structural characterization of a novel xylolipid biosurfactant from Lactococcus lactis and analysis of antibacterial activity against multi-drug resistant pathogens. Bioresour Technol 101(22):8851–8854

Sánchez M, Teruel JA, Espuny MJ, Marqués A, Aranda FJ, Manresa Á, Ortiz A (2006) Modulation of the physical properties of dielaidoylphosphatidylethanolamine membranes by a dirhamnolipid biosurfactant produced by *Pseudomonas aeruginosa*. Chem Phys Lipids 142 (1):118–127

Sauvageau J, Ryan J, Lagutin K, Sims IM, Stocker BL, Timmer MS (2012) Isolation and structural characterisation of the major glycolipids from *Lactobacillus* plantarum. Carbohydr Res 357:151–156

Shabtai YOSSEF, Gutnick DL (1985) Exocellular esterase and emulsan release from the cell surface of *Acinetobacter calcoaceticus*. J Bacteriol 161(3):1176–1181

Sharma D, Saharan BS, Chauhan N, Procha S, Lal S (2015a) Isolation and functional characterization of novel biosurfactant produced by Enterococcus faecium. SpringerPlus 4(1):4

Sharma D, Saharan BS, Chauhan N, Procha S, Lal S (2015b) Isolation and functional characterization of novel biosurfactant produced by Enterococcus faecium. SpringerPlus 4(1):4

Sharma D, Saharan BS (2014) Simultaneous Production of biosurfactants and bacteriocins by probiotic *Lactobacillus casei* MRTL3. Int J Microbiol 2014

Sharma D, Saharan BS, Chauhan N, Bansal A, Procha S (2014) Production and structural characterization of *Lactobacillus helveticus* derived biosurfactant. Sci World J 2014

Singh M, Saini VS, Adhikari DK, Desai JD, Sista VR (1990) Production of bioemulsifier by a
    SCP-producing strain of *Candida tropicalis* during hydrocarbon fermentation. Biotechnol Lett
    12(10):743–746
Syldatk C, Lang S, Wagner F, Wray V, Witte L (1985) Chemical and physical characterization of
    four interfacial-active rhamnolipids from Pseudomonas spec. DSM 2874 grown on
    n-alkanes. Zeitschrift für Naturforschung C 40(1–2):51–60
Tahmourespour A, Salehi R, Kermanshahi RK (2011) *Lactobacillus* acidophilus-derived
    biosurfactant effect on gtfB and gtfC expression level in Streptococcus mutans biofilm cells.
    Braz J Microbiol 42(1):330–339
Thavasi R, Jayalakshmi S, Banat IM (2011) Application of biosurfactant produced from peanut oil
    cake by *Lactobacillus* delbrueckii in biodegradation of crude oil. Bioresour Technol 102
    (3):3366–3372
Thavasi R, Jayalakshmi S, Balasubramanian T, Banat IM (2008) Production and characterization
    of a glycolipid biosurfactant from *Bacillus megaterium* using economically cheaper sources.
    World J Microbiol Biot 24(7):917–925
Toribio J, Escalante AE, Caballero-Mellado J, González-González A, Zavala S, Souza V,
    Soberón-Chávez G (2011) Characterization of a novel biosurfactant producing *Pseudomonas*
    koreensis lineage that is endemic to Cuatro Ciénegas Basin. Syst Appl Microbiol 34
    (7):531–535
Van Hoogmoed CG, Van der Mei HC, Busscher HJ (2004) The influence of biosurfactants
    released by S. *mitis* BMS on the adhesion of pioneer strains and cariogenic bacteria. Biofouling
    20(6):261–267
Vecino X, Barbosa-Pereira L, Devesa-Rey R, Cruz JM, Moldes AB (2015) Optimization of
    extraction conditions and fatty acid characterization of *Lactobacillus* pentosus cell-bound
    biosurfactant/bioemulsifier. J Sci Food Agric 95(2):313–320
Vecino X, Devesa-Rey R, Moldes AB, Cruz JM (2014) Formulation of an alginate-vineyard
    pruning waste composite as a new eco-friendly adsorbent to remove micronutrients from
    agroindustrial effluents. Chemosphere 111: 24–31
Velraeds MM, Van der Mei HC, Reid G, Busscher HJ (1996) Inhibition of initial adhesion of
    uropathogenic Enterococcus faecalis by biosurfactants from *Lactobacillus* isolates. Appl
    Environ Microbiol 62(6):1958–1963

Chapter 5
Substrates and Production of Biosurfactants

Abstract Inexpensive industrial scale production for biosurfactants remains a problematic issue. Development in production measures has facilitated to some magnitude and can lead to additional advances. The utilization of the unconventional substrates like agro-industrial liquid and solid wastes is the efficient approaches for cost-effective biosurfactants production. Additional strategy involves exhausting low-cost substrates with insignificant or no worth. The direction to achieve pure biosurfactants is largely reliant on various extraction and purification stages. Utilization of simple substrates with low-cost product recovery approaches will cut down the economy of fermentation process and the utilization of agro-industrial substrates signifies an encouraging step to achieve that aim of the low-cost productions.

Keywords Agro-industrial waste · Whey · Fermentation · Lignocellulosic waste

Introduction

The production budget of any product of microbial origin is directed by these simple aspects: (i) preliminary cost of raw material, (ii) convenience of appropriate and low-cost production and downstream processing, and (iii) product yield. Therefore, for economic restrictions associated with commercial production of biosurfactants, these basic approaches are implemented globally to make biosurfactant production more economically viable (i) utilization of inexpensive waste substrates to reduce the raw substrate expenses; (ii) improvement in bioprocesses, optimization of growth medium and growth conditions, downstream processing, and purification; and (iii) selection/construction of hyper-producing mutants or genetically engineered strains for improved biosurfactant yields. Inexpensive industrial scale production for biosurfactants remains a problematic issue. Development in production measures has facilitated to some magnitude and can lead to additional advances (Mukherjee et al. 2006). Walter et al. 2010 stated that the causes for inadequate use of biosurfactants in industry are the selection of

© The Author(s) 2016
D. Sharma et al., *Biosurfactants of Lactic Acid Bacteria*,
SpringerBriefs in Microbiology, DOI 10.1007/978-3-319-26215-4_5

Table 5.1 Substrates reported for LAB-derived biosurfactants

S. no.	LAB strain	Substrate used	Study
1.	*Lactobacillus* spp.	MRS broth	Velraeds et al. (1996)
2.	*Streptococci thermophilus*	M-17 broth	Busscher et al. (1997)
3.	*Streptococci mitis*	–	Van Hoogmoed et al. (2004)
4.	*Lactobacillus fermentum* RC-14	MRS broth	Heinemann et al. (2000)
5.	*Lactococcus lactis* 53	MRS broth	Rodrigues et al. (2004)
6.	*Lactobacillus casei* CECT 525, *Lactobacillus rhamnosus* CECT 288, *Lactobacillus pentosus* CECT 4023 *and Lactobacillus coryniformis* subsp. *torquens* CECT 25600	MRS broth	Rodrigues et al. (2006a, b)
7.	*Streptococcus thermophiles* A	M17 broth	Rodrigues et al. (2006a, b)
8.	*Lactobacillus pentosus*	Grape marc	Rivera et al. (2007)
9.	*Lactobacillus pentosus*	Hemicellulosic hydrolysate	Moldes et al. (2007)
10.	*Lactobacillus acidophillus*	MRS broth	Walencka et al. (2008)
11.	*Lactobacillus acidophillus*	Vine-trimming waste	Portilla et al. (2008)
12.	*Lactococcus lactis*	MRS broth	Saravanakumari and Mani (2010)
13.	*Lactobacillus paracasei*	MRS broth	Gudiña et al. (2010a, b)
14.	*Lactococcus lactis*	MRS broth	Rodríguez et al. (2010)
15.	*Lactococcus paracasei* subsp. *Paracasei* A20	MRS broth	Gudiña et al. (2010a, b)
16.	*Lactobacillus delbrueckii* sbusp. *delbruckii*	MRS broth	Fracchia et al. (2010)
17.	*Lactobacillus delbrueckii*	Peanut oil cake	Thavasi et al. (2011a)
18.	*Lactobacillus acidophilus*	MRS broth	Tahmourespour and Kermanshahi (2011)
19.	*Lactobacillus fermentii* and *Lactobacillus rhamnosus*	MRS broth	Brzozowski et al. (2011)
20.	*Lactobacillus* spp.	MRS broth	Kermanshahi and Peymanfar (2012)
21.	*Lactobacillus plantarum*	MRS broth	Sauvageau et al. (2012)
22.	*Lactobacillus reutri*	MRS broth	Salehi et al. (2014)
23.	*Lactobacillus* spp.	MRS broth	Augustin and Tene Hippolyte (2012)

(continued)

Table 5.1 (continued)

S. no.	LAB strain	Substrate used	Study
24.	*Lactobacillus pentosus* and *Lactobacillus plantarum* co-culture	Hydrolysates of trimming waste	Rodríguez-Pazo et al. (2013)
25.	*Lactobacillus* spp.	MRS broth	Gomaa et al. (2013)
26.	*Lactobacillus pentosus*	Hemicellulosic sugars	Vecino et al. (2013)
27.	*Lactobacillus plantarum* CFR2194	MRS broth	Madhu and Prapulla (2014)
28.	*Lactobacillus pentosus*	Hemicellulose sugars	Vecino et al. (2014a)
29.	*Lactobacillus pentosus*	Corn steep liquor	Vecino et al. (2014b)
30.	*Lactobacillus brevis*	MRS broth	Ceresa et al. (2015)
31.	*Lactobacillus jensenii* and *Lactobacillus rhamnosus*	MRS broth	Sambanthamoorthy et al. (2014)

expensive materials as substrates, inadequate product yields, and production of mixtures in place of pure product. The utilization of the unconventional substrates like agro-industrial liquid and solid wastes is the efficient approaches for cost-effective biosurfactants production. Additional strategy involves exhausting low-cost substrates with insignificant or no worth by designing of processes which utilize inexpensive substrates or industrial pollutants. Although this seems simple, the key problem related with the current strategy is the assortment of efficient waste substrate with the proper balance of nutrients that allow microbial growth and product formation. A number of unconventional substrates are presently accessible as nutrients for commercial fermentations process, specifically agricultural and food processing by-products (Table 5.1) (Da Silva et al. 2009; Saharan et al. 2011; Savarino et al. 2007). The expansion of low-budget routes and inexpensive substrates can account for 10–30 % of product developed (Mukherjee et al. 2006; Mutalik et al. 2008). Smyth et al. (2010) discussed that stress should be on inexpensive and simple downstream processing approaches. The direction to achieve pure biosurfactants is largely reliant on various extraction and purification stages. Utilization of simple substrates with low-cost product recovery approaches will cut down the economy of fermentation process and the utilization of agro-industrial substrates signifies an encouraging step to achieve that aim of low-cost productions (Mukherjee et al. 2006).

Practice of statistical approaches including various tools like factorial design and response surface methodology (RSM) will benefit in efficient media and growth optimization for production of biosurfactants. The convenience of process development with narrow downstream processing will provide noteworthy cost-effective returns.

Agro-Industrial Substrates

Producing of biosurfactants from various agro-industrial wastes is a viable and promising route for low-cost bioprocess (Cameotra and Makkar 2010; Moldes et al. 2007). Wastes originated from metropolitan activities and wastes obtained from agro-industrial preparations are vital source of various lignocellulosic wastes (Moldes et al. 2007). The huge potential related with these substrates has not been valorized properly. These low-cost agro-industrial substrates used for the production of biosurfactants from LAB include hemicellulosic hydrolysates, distillery, peanut cakes, corn steep liquor, grape marc, and cheese whey (Rivera et al. 2007; Moldes et al. 2007; Portilla et al. 2008; Rodríguez-Pazo et al. 2013; Vecino et al. 2014a, b). These agro-industrial substrates are certain examples of wastes that can be utilized as a feedstock for large-scale production of biosurfactants. The valorization of such kind of substrates plays a dual role, of producing a functional product and simultaneously sinking waste dumping problem.

Biosurfactant Production from Cheese Whey

The dairy industry has significant volume of various waste by-products like cheese whey. Cheese whey is a liquid leftover of cheese processing, rich in lactose content (approximately 75 % of dry matter) and also comprising other water-soluble constituents (approximately 12–14 % protein). The world whey collection is above 160 million tons per year, with a progressive 1–2 % annual growth rate (OECD-FAO 2008; Smithers 2008). Cheese whey recognized with high amount of biochemical oxygen demand (BOD) of about 30–50 g L^{-1} and a chemical oxygen demand (COD) of about 60–80 g L^{-1}. Lactose moieties are fundamentally accountable for the high BOD and COD content of cheese whey (Ghaly and Kamal 2004; Kisaalita et al. 1990). Whey has a high BOD and COD value and its removal can be challenging exclusively for countries depending on dairy economy. However, nearly half of the global cheese whey obtained is not properly treated and is being thrown away as untreated effluent to local water bodies.

Thus, cheese whey illustrates as a vital environmental problem because of the high amount generated and its BOD and COD exhibition (Marwaha and Kennedy 1988; Mawson 1994). Therefore, the disposal of cheese whey is an expensive and problematic issue as concerned to the environment for dairy processing sector. The valorization of whey has been a challenge for dairy industries in the production of cheese around the globe. With increasing global cheese consumption and huge demand, the amount of generated cheese whey also raised up (Rajeshwari et al. 2000; De Wit 2001).

Daniel et al. (1998) observed that the hyper yields of sophorolipids (glycolipid) production were reported with cheese whey concentrate and rapeseed oil as low-cost substrate. Daverey and Pakshirajan (2010) Sophorolipids production by

Candida bombicola has been reported by Pakshirajan (2010) on growth medium comprising mixed hydrophilic material with yeast extract and oleic acid.

Lactobacillus helveticus MRTL91 was found to be a potent biosurfactant-producing probiotic microorganism utilizing cheese whey as alternative nutrient source. The highest reduction in surface tension of whey-based medium was achieved after 10 h of fermentation, i.e., 39.5 m Nm^{-1} (Sharma et al. 2014). Biosurfactant production from *L. helveticus* was found to be growth associated at lab-scale batch fermentation. The pH of the cheese whey-based medium was controlled at pH 6.2 throughout the fermentation and that positively contributed for hyper biomass production with maximal lactose utilization within 10 h after inoculation. The lactose present in the whey was exhausted in first 24 h, and further incubation results in cell death. Biosurfactant concentration was found to be maximal, i.e., 0.80 gL^{-1}. Increase in lactose concentration into the fermentative medium yields higher biomass and higher biosurfactant obtained from different lactobacilli (Gudiña et al. 2012). The results achieved for *L. helveticus* MRTL91 established that the strain is a potent biosurfactant producer and cheese whey-based fermentative medium can be used as an alternative nutrient for large-scale production of microbial surfactants.

Rodrigues et al. (2006a, b) reported the production process kinetics of surface active agents by *L. pentosus* using cheese whey as an alternative substrate. The minimal value of surface tension reduction was obtained in the stationary growth phase of the microbial cell growth, i.e., 45 mN/m, and the total decrease in the surface tension surpassed 8 mN/m (Velraeds et al. 1996). The surface tension reduction was equated with the surface tension of whey-based medium control (54 mN/m). Fermentation by *L. pentosus* was carried out utilizing cheese whey as culture medium for *L. pentosus*.

Cheese whey was only used as the medium; nothing else in addition to whey has been used for media preparation. Lactose utilization at various times of fermentation turns out to be constant; however, microbial biomass is still increasing probably *L. pentosus* consuming additional medium constituents rather than lactose present in the whey. So a large portion of the residual sugar content was observed at the end of fermentation cycle. Equating the kinetic constraints attained with both the medium (MRS broth and Cheese whey-based medium), it was probable to notice that a lower value of μmax (10 % less than with MRS medium) was achieved with cheese whey-based medium with lower X_{max}, i.e., about one-third of the value achieved with MRS medium. Further, culture medium and cultural conditions optimization possibly will be achieved by hyper biosurfactant concentrations from probiotic organisms.

In another study, Rodrigues et al. (2006a, b) reported the low-cost production of biosurfactants from probiotic *Lactococcus lactis* using cheese whey and molasses-based media. Several models have been developed to define the response of the trials concerning to sugar utilization especially lactose consumption, cell biomass, and biosurfactant production. In the present case, MRS and M17 broth were used as control trials. When the MRS (synthetic media) was substituted by cost-effective media, as dairy whey, fermentations were conceded out efficiently with high productivities of biosurfactant. An upsurge of approximately 1.2–1.5

times in the yield of produced biosurfactant per gram cell dry weight and 60–80 % medium formulation expenses saving were achieved. The results achieved reflected that supplemented cheese whey medium can be utilized as a comparatively cheaper and reasonable substitute to MRS medium for biosurfactant production by probiotic lactic acid bacteria.

Biosurfactant Production from Lignocellulosic Waste

Lignocellulosic waste is among the most ample organic carbon existing on planet and they are the main constituents of various wastes obtained from different industries, agricultures, and cities (Kukhar 2009). Lignocellulose substrates comprise predominantly three kinds of biopolymers like cellulose, hemicellulose, and lignin that are intensely interconnected and bonded chemically by non-covalent and covalent cross-linkages. Microbial deprivation of these bio-macromolecules by various microorganisms such as fungi and bacteria has been broadly reported. Generally, they have been used as low-cost substrates for the production of bio-ethanol and certain organic acids (Taherzadeh and Karimi 2007).

As an inexpensive substrate, lignocellulosic opulent agricultural remains can be utilized for producing significant microbial metabolites like biosurfactants. There are certain reports regarding valorization of the lignocellulosic substrates for the production of microbial surfactant. Utilization of lignocellulosic substrates for the production of organic acid, i.e., lactic acid, has been reported with various LAB (Bustos 2007). Simultaneous production of lactic acid and biosurfactant has been reported with *Lactobacillus* sp. using hemicellulosic hydrolysates from different agricultural remains. Simultaneous production approaches make biosurfactant further cost effective. It has been concluded from the above study that hemicellulosic sugars from the agricultural waste are motivating substrates for the economical production of biosurfactants.

Distilled grape marc (by product of wine industry) a kind of lignocellulosic substrate has been valorized for the production of biosurfactant using LAB (Portilla-Rivera et al. 2008). A huge amount of grape marc has been accumulated in viticulture which produced after pressing the crushing grapes in the course of wine fermentation. Distilled grape marc residues left unutilized which has vast quantity of hemicellulosic and various organic acids that can be exploited for the production of lactic acid and biosurfactants (Rivera et al. 2007). They observed a yield of 4.8 mg/l of cell-bound biosurfactants per gram of sugar utilized by *L. pentosus*. Acid hydrolysis of grape marc was conceded out using acid pretreatment and 130 ° C, in direction to achieve monomeric sugars. Monomeric sugar was supplemented with corn steep liquor and yeast extract at a concentration of 10 g/L tzo perform fermentation. Xylose was the chief monomeric sugar obtained trailed by glucose and arabinose sugar. Moreover, *L. pentosus* produced 4.8 mg/L of cell-bound biosurfactants per gram of sugars utilized. Present study establishes the prospect of utilizing cost-effective agricultural waste as substrates for biosurfactant production.

Agricultural residues (bran, vine shoots, corn cobs, and *E. globulus* chips) for concurrent production of lactic acid and biosurfactant were derived from *Lactobacillus pentosus* after nutrient supplementation (Moldes et al. 2007). Biosurfactants derived from *L. pentosus* were measured by captivating the reduction in surface tension. The maximal reduction in surface tension, i.e., 21.3 units, was observed utilizing hemicellulosic hydrolyzates derived from trimming vine shoots containing 0.71 g of biosurfactant per g of cell mass and 25.6 g of lactic acid/L, respectively. On the contrary, barley bran hydrolyzates simply produced 0.28 g of biosurfactant per g of cell mass and 33.2 g of lactic acid/L. During the fermentation of hemicellulosic hydrolyzates, bacteria produced lactic acid and cell-bound biosurfactants. Out of the lignocellulosic material evaluated, trimming vine shoots gave the maximal amount of biosurfactants while barley bran produced the maximal amount of lactic acid amount. The usefulness of vine-trimming shoots for carrying out fermentation for the production of biosurfactants may perhaps be associated with the maximal amount of glucose in the hydrolysates.

One more motivating tactic for achieving more productive outcomes will be co-production of surface active agents and other significant microbial metabolites. Various researches on such co-production of metabolites, e.g., bio-plastic production, lactic acid, biosurfactant, lactic acid, bacteriocin, and biosurfactants, have been documented and performed (Sharma et al. 2014; Rodríguez-Pazo et al. 2013). This approach fits together in cooperation with the emerging attention of users toward the management of wastes and economical benefits. Rodríguez-Pazo et al. (2013) valorized cellulosic and hemicellulosic residues of vine shoots for the production of biosurfactants, lactic acid, phenyl, and lactic acid obtained from *L. plantarum* and *L. pentosus*. The present study equates the study of the prospective of hemicellulosic and cellulosic residues of vine shoots as an inexpensive carbon source for lactic acid and biosurfactant production by individual or co-culture of both.

Various metabolites like lactic acid, phenyl lactic acid, and biosurfactants can be derived by *L. plantarum* and *L. pentosus* in self-regulating cultures or simultaneous fermentations utilizing hemicellulosic portions of vine shoots as a sole carbon source. Although all the approaches evaluated proved to be effective, the simultaneous production by both microorganisms improved the amount of lactic acid and phenyl lactic and also augmented the decrease of surface tension. Subsequently, it can be established that this strategy facilitates the synergistic use of metabolic pathways of the microorganisms with the added advantage of directing the bio-process in one step.

Biosurfactant from Synthetic Medium

Biosurfactant derived from different lactic acid bacteria can be produced by chemically defined medium. According to Gudiña et al. (2010a, b), *Lactobacillus paracasei* ssp. *paracasei* A20 isolated from a dairy processing plant could be used for biosurfactant production using MRS broth medium with a reduction in the surface tension of 6.4 and 22.0 mN/m, separately. Biosurfactant derived from *L. paracasei* ssp. *paracasei* A20 was evaluated in various culture broth compositions. The amendment of various nitrogen sources showed that yeast extract is vital for bacterial biomass, whereas peptone is critical for biosurfactant production. The amendment of peptone and yeast extract yielded highest production when equated to the typical MRS medium, with a surface tension decrease of 24.5 mN/m. However, when nitrogen sources are replaced by the ammonium, there is no growth, possibly due to the deficiency of vital nutrients. LAB are exceptionally demanding adapted to complex nutrient. Lactic acid bacteria not only require carbon source as energy but also require nucleotides, amino acids, and vitamins for growth, due to the nonexistence of biosynthetic pathways. The results observed with nutrient media that composed of two nitrogen sources noticeably showed that amendment of yeast extract is the vital parameter for bacterial growth and fall behind by meat extract. As for surfactant production, it appears that peptone is the vital constituent. The maximal cell-bound biosurfactant amounts were obtained using media composed of peptone.

Other Miscellaneous Substrates

The above-conferred substrates are utilized to reduce the cost to cut down the overall biosurfactant production developments. Various different substrates have been reported by several authors with respect to the biosurfactant obtained from different LAB strains (Table 5.1). Thavasi et al. (2011b) established the fact that peanut oil cake can be used as a low-cost potent substrate for the production of biosurfactant from *L. delbrueckii*. Peanut oil cake is rich in carbohydrates, lipid, and obtained during peanut oil pressing. In progression of biosurfactant production with peanut oil cake utilized as a low-cost substrate, maximal biosurfactant obtained was 5.35 mg/L after 144 h of incubation under stationary phase of growth with maximal biomass of 9.04 log CFU ml^{-1}. This showed the prospective of peanut cake as an inexpensive substrate for biosurfactant production by LAB.

Conclusion

Biosurfactants are vital metabolites in understanding of the amount sold and of their immense applications. Economic improvements in lieu with the operational and production expenses and applicability will be the decisive factors for utilization of low-cost substrates for the production of biosurfactants. New prospects will result from the marked applications of biosurfactants in biological applications as antimicrobial agents. The application of cost-effective production approaches and utilization of waste as a substrate is also attaining ground. The real implication of these kinds of approaches will be defensible only when these technologies will be commercially viable. Prosperous scale up of any biotechnological product will be influenced mainly on its bio-economics. In current scenario, the costs of biosurfactants production are not viable as comparable to the chemical surfactants due to high production costs. However, the utilization of low-cost substrates and optimized cultural conditions together with efficient product recovery and hyper-producing strains can make process economically feasible. A sensible and operational blend of these approaches might, in the coming future, lead the process development in the direction of large-scale economically viable production of microbial surfactants.

References

Augustin M, Tene Hippolyte M (2012). Screening of biosurfactants properties of cell-free supernatants of cultures of *Lactobacillus* spp. isolated from a local fermented milk (Pendidam) of Ngaoundere (Cameroon). Int J Eng Res Appl 2(5):974–985

Brzozowski B, Bednarski W, Golek P (2011) The adhesive capability of two *Lactobacillus* strains and physicochemical properties of their synthesized biosurfactants. Food Technol Biotechnol 49(2):177

Busscher HJ, Van Hoogmoed CG, Geertsema-Doornbusch GI, Van der Kuijl-Booij M, Van der Mei HC (1997) Streptococcus thermophilus and its biosurfactants inhibit adhesion by *Candida* spp. on silicone rubber. Appl Environ Microbiol 63(10):3810–3817

Bustos G, De la Torre N, Moldes AB, Cruz JM, Domínguez JM (2007) Revalorization of hemicellulosic trimming vine shoots hydrolyzates trough continuous production of lactic acid and biosurfactants by *L. pentosus*. J Food Eng 78(2):405–412

Cameotra SS, Makkar RS (2010) Biosurfactant-enhanced bioremediation of hydrophobic pollutants. Pure Appl Chem 82(1):97–116

Ceresa C, Tessarolo F, Caola I, Nollo G, Cavallo M, Rinaldi M, Fracchia L (2015) Inhibition of Candida albicans adhesion on medical-grade silicone by a *Lactobacillus*-derived biosurfactant. J Appl Microbiol 118(5):1116–1125

Da Silva GP, Mack M, Contiero J (2009) Glycerol: a promising and abundant carbon source for industrial microbiology. Biotechnol Adv 27(1):30–39

Daniel HJ, Reuss M, Syldatk C (1998) Production of sophorolipids in high concentration from deproteinized whey and rapeseed oil in a two stage fed batch process using Candida bombicola ATCC 22214 and Cryptococcus curvatus ATCC 20509. Biotechnol Lett 20(12):1153–1156

Daverey A, Pakshirajan K (2010) Sophorolipids from Candida bombicola using mixed hydrophilic substrates: production, purification and characterization. Colloids Surf B 79(1):246–253

De Wit JN (2001) Lecturer's handbook on whey and whey products. European whey products association, Brussels

Fracchia L, Cavallo M, Allegrone G, Martinotti MG (2010) A *Lactobacillus*-derived biosurfactant inhibits biofilm formation of human pathogenic Candida albicans biofilm producers. Appl Microbiol Biotechnol 2:827–837

Ghaly AE, Kamal MA (2004) Submerged yeast fermentation of acid cheese whey for protein production and pollution potential reduction. Water Res, 38(3):631–644

Golek P, Bednarski W, Brzozowski B, Dziuba B (2009) The obtaining and properties of biosurfactants synthesized by bacteria of the genus *Lactobacillus*. Ann Microbiol 59(1):119–126

Gomaa EZ (2013) Antimicrobial activity of a biosurfactant produced by Bacillus licheniformis strain M104 grown on whey. Braz Archives Biol Technol 56(2):259–268

Gudiña EJ, Teixeira JA, Rodrigues LR (2010a) Isolation and functional characterization of a biosurfactant produced by *Lactobacillus paracasei*. Colloids Surf B 76(1):298–304

Gudiña EJ, Pereira JF, Rodrigues LR, Coutinho JA, Teixeira JA. (2012) Isolation and study of microorganisms from oil samples for application in microbial enhanced oil recovery. Int Biodeterior Biodegradation 68:56–64

Gudiña EJ, Rocha V, Teixeira JA, Rodrigues LR (2010b) Antimicrobial and antiadhesive properties of a biosurfactant isolated from *Lactobacillus paracasei* ssp. *paracasei* A20. Lett Appl Microbiol 50(4):419–424

Heinemann C, van Hylckama Vlieg JE, Janssen DB, Busscher HJ, van der Mei HC, Reid G (2000) Purification and characterization of a surface-binding protein from *Lactobacillus fermentum* RC-14 that inhibits adhesion of *Enterococcus faecalis* 1131. FEMS Microbiol Lett 190 (1):177–180

Kermanshahi RK, Peymanfar S (2012) Isolation and identification of lactobacilli from cheese, yoghurt and silage by 16S rDNA gene and study of bacteriocin and biosurfactant production. Jundishapur J Microbiol 5(4):528–532

Kisaalita WS, Lo KV, Pinder KL (1990) Influence of whey protein on continuous acidogenic degradation of lactose. Biotechnol Bioeng 36(6):642–646

Kukhar VP (2009) Biomass–Renewable feedstock for organic chemicals (White Chemistry)

Mawson AJ (1994) Bioconversions for whey utilization and waste abatement. Bioresour Technol, 47(3):195–203

Madhu AN, Prapulla SG (2014) Evaluation and functional characterization of a biosurfactant produced by *Lactobacillus plantarum* CFR 2194. Appl Biochem Biotechnol 172(4):1777–1789

Marwaha SS, Kennedy JF (1988) Whey—pollution problem and potential utilization. Int J Food Sci Technol 23(4):323–336

Moldes AB, Torrado AM, Barral MT, Domínguez JM (2007) Evaluation of biosurfactant production from various agricultural residues by *Lactobacillus pentosus*. J Agric Food Chem 55(11):4481–4486

Mukherjee S, Das P, Sen R (2006) Towards commercial production of microbial surfactants. Trends Biotechnol 24(11):509–515

Mutalik SR, Vaidya BK, Joshi RM, Desai KM, Nene SN (2008) Use of response surface optimization for the production of biosurfactant from *Rhodococcus* spp. MTCC 2574. Bioresour Technol 99(16):7875–7880

Portilla OM, Rivas B, Torrado A, Moldes AB, Domínguez JM (2008) Revalorisation of vine trimming wastes using *Lactobacillus acidophilus* and *Debaryomyces hansenii*. J Sci Food Agric 88(13):2298–2308

Rajeshwari KV, Balakrishnan M, Kansal A, Lata K, Kishore VVN (2000) State-of-the-art of anaerobic digestion technology for industrial wastewater treatment. Renew Sustain Energy Rev 4(2):135–156

Rivera OMP, Moldes AB, Torrado AM, Domínguez JM (2007) Lactic acid and biosurfactants production from hydrolyzed distilled grape marc. Process Biochem 42(6):1010–1020

Rivera OMP, Torrado A., Domínguez JM, Moldes AB (2008) Stability and emulsifying capacity of biosurfactants obtained from lignocellulosic sources using *Lactobacillus pentosus*. J Agr Food Chem 56(17):8074–8080

Rivera OMP, Martínez GA, Enríquez LJ, Landaverde PAV, González JMD (2015) Lactic acid and biosurfactants production from residual cellulose films. Appl Biochem Biotechnol 177 (5):1099–1114

Rodrigues L, Van der Mei H, Teixeira JA, Oliveira R (2004) Biosurfactant from Lactococcus lactis 53 inhibits microbial adhesion on silicone rubber. Appl Microbiol Biotechnol 66(3):306–311

Rodrigues LR, Teixeira JA, Oliveira R (2006a) Low-cost fermentative medium for biosurfactant production by probiotic bacteria. Biochem Eng J 32(3):135–142

Rodrigues L, Banat IM, Teixeira J, Oliveira R (2006b) Biosurfactants: potential applications in medicine. J Antimicrob Chemother 57(4):609–618

Rodríguez N, Salgado JM, Cortés S, Domínguez JM (2010) Alternatives for biosurfactants and bacteriocins extraction from *Lactococcus lactis* cultures produced under different pH conditions. Lett Appl Microbiol 51(2):226–233

Rodríguez-Pazo N, Salgado JM, Cortés-Diéguez S, Domínguez JM (2013) Biotechnological production of phenyllactic acid and biosurfactants from trimming vine shoot hydrolyzates by microbial coculture fermentation. Appl Biochem Biotechnol 169(7):2175–2188

Saharan BS, Sahu RK, Sharma D (2011) A review on biosurfactants: fermentation, current developments and perspectives. Gen Eng Biotechnol J 2011(1):1–14

Salehi R, Savabi O, Kazemi M (2014) Effects of *Lactobacillus* reuteri-derived biosurfactant on the gene expression profile of essential adhesion genes (gtfB, gtfC and ftf) of *Streptococcus mutans*. Adv Biomed Res 3

Sambanthamoorthy K, Feng X, Patel R, Patel S, Paranavitana C (2014) Antimicrobial and antibiofilm potential of biosurfactants isolated from lactobacilli against multi-drug-resistant pathogens. BMC Microbiol 14(1):197

Saravanakumari P, Mani K (2010) Structural characterization of a novel xylolipid biosurfactant from *Lactococcus lactis* and analysis of antibacterial activity against multi-drug resistant pathogens. Bioresour Technol 101(22):8851–8854

Sauvageau J, Ryan J, Lagutin K, Sims IM, Stocker BL, Timmer MS (2012) Isolation and structural characterisation of the major glycolipids from *Lactobacillus plantarum*. Carbohydr Res 357:151–156

Savarino P, Montoneri E, Biasizzo M, Quagliotto P, Viscardi G, Boffa V (2007) Upgrading biomass wastes in chemical technology. Humic acid-like matter isolated from compost as chemical auxiliary for textile dyeing. J Chem Technol Biotechnol 82(10):939–948

Sharma D, Saharan BS (2014) Simultaneous production of biosurfactants and bacteriocins by probiotic *Lactobacillus casei* MRTL3. Int J Microbiol 2014

Sharma D, Saharan BS, Chauhan N, Bansal A, Procha S (2014) Production and structural characterization of *Lactobacillus helveticus* derived biosurfactant. Sci World J 2014

Smyth TJP, Perfumo A, McClean S, Marchant R, Banat IM (2010) Isolation and analysis of lipopeptides and high molecular weight biosurfactants. In: Handbook of hydrocarbon and lipid microbiology, Springer, Berlin Heidelberg, pp 3687–3704

Taherzadeh MJ, Karimi K (2007) Acid-based hydrolysis processes for ethanol from lignocellulosic materials: a review. BioResour 2(3):472–499

Tahmourespour A, Kermanshahi RK (2011) The effect of a probiotic strain (*Lactobacillus acidophilus*) on the plaque formation of oral Streptococci. Bosnian J Basic Med Sci/Udruzenje basicnih mediciniskih znanosti = Association of Basic Medical Sciences 11(1):37–40

Tahmourespour A., Salehi R, Kermanshahi RK, Eslami G (2011) The anti-biofouling effect of Lactobacillus fermentum-derived biosurfactant against *Streptococcus mutans*. Biofouling 27 (4):385–392

Thavasi R, Jayalakshmi S, Banat IM (2011a) Application of biosurfactant produced from peanut oil cake by Lactobacillus delbrueckii in biodegradation of crude oil. Bioresour Technol 102 (3):3366–3372

Thavasi R, Jayalakshmi S, Banat IM (2011b) Effect of biosurfactant and fertilizer on biodegradation of crude oil by marine isolates of *Bacillus megaterium, Corynebacterium kutscheri and Pseudomonas aeruginosa*. Bioresour Technol 102(2):772–778

van Hoogmoed CG, van der Mei HC, Busscher HJ (2004) The influence of biosurfactants released by *S. mitis* BMS on the adhesion of pioneer strains and cariogenic bacteria. Biofouling 20 (6):261–267

Vecino X, Devesa-Rey R, Cruz JM, Moldes AB (2013) Evaluation of biosurfactant obtained from *Lactobacillus pentosus* as foaming agent in froth flotation. J Environ Manage 128:655–660

Vecino X, Barbosa-Pereira L, Devesa-Rey R, Cruz JM, Moldes AB (2014a) Study of the surfactant properties of aqueous stream from the corn milling industry. J Agric Food Chem 62 (24):5451–5457

Vecino X, Devesa-Rey R, Moldes AB, Cruz JM (2014b) Formulation of an alginate-vineyard pruning waste composite as a new eco-friendly adsorbent to remove micronutrients from agroindustrial effluents. Chemosphere 111:24–31

Velraeds MM, Van der Mei HC, Reid G, Busscher HJ (1996) Inhibition of initial adhesion of uropathogenic Enterococcus faecalis by biosurfactants from *Lactobacillus* isolates. Appl Environ Microbiol 62(6):1958–1963

Walencka E, Różalska S, Sadowska B, Różalska B (2008) The influence of *Lactobacillus acidophilus*-derived surfactants on staphylococcal adhesion and biofilm formation. Folia Microbiol 53(1):61–66

Walter V, Syldatk C, Hausmann R (2010) Screening concepts for the isolation of biosurfactant producing microorganisms. In: Biosurfactants. Springer, New York, pp 1–13

Chapter 6
Applications of Biosurfactants

Abstract Biosurfactants are superior to chemical surfactants, because of their biological origin, biodegradability, and low level of toxicity. For this reason biosurfactants have been extensively considered for application in food processing, cosmetics formulations, microbial enhanced oil recovery, soil washing, emulsifying materials, and bioremediation agents. Of all the activities of microbial biosurfactants, antimicrobial and anti-adhesive properties against various pathogens and their probiotic behavior are the most significant properties for therapeutics applications. In addition to the above discussed applications of biosurfactants, there are several other sectors to explore biosurfactants such as food and feed processing, active food ingredients, and green surface cleaning agents for food and biomedical surfaces.

Keywords Antimicrobial · Antibiofilm · Surgical equipment · Silicone catheters · Food

Introduction

Microbial surfactants are a collection of structurally different molecules obtained from various microorganisms and are chiefly categorized on the basis of their chemical composition and producing microorganisms. Generally, they are composed of a hydrophilic moiety and a hydrophobic moiety. Biosurfactants can be generally distributed into two classes: low-molecular-weight molecules termed biosurfactants, like glycolipids, and high-molecular-weight biosurfactants or bioemulsifiers (Neu 1996; Rosenberg and Ron 1997; Smyth et al. 2010). The importance of biosurfactants has progressively increased in the course of the past decade. Biosurfactants are superior to chemical surfactants because of their biological origin, biodegradability, and low level of toxicity. For this reason biosurfactants have been extensively considered for application in food processing, cosmetics formulations, microbial enhanced oil recovery, soil washing, emulsifying materials, and bioremediation agents (Marchant and Banat 2012). Biosurfactants can alter the microbial adhesion as they accumulate at interfaces of two fluid phases (Van Hamme et al. 2006). Similarly, biosurfactants can disrupt plasma membranes that result in increased cell membrane permeability and, eventually, seepage of cytoplasmic content of the cell (Bharali et al. 2013). Of all the activities of microbial biosurfactants, antimicrobial and antiadhesive properties against various pathogens and their probiotic behavior are the most significant properties for therapeutic applications (Sharma et al. 2015; Marchant and Banat 2012; Van Hamme et al. 2006). Various microbial surfactants have been described as appropriate substitutions to synthetic drugs and antimicrobials.

Lactic acid bacteria (LAB) characterize an important part of the human gastrointestinal and genitourinary tracts. LAB are regarded as probiotic agents that inhibit the growth of various pathogens by secreting various compounds (e.g., lactic acid, hydrogen peroxide, bacteriocins, and biosurfactants).

This chapter discusses the existing potential of biosurfactants used as therapeutic application. We aim to make available new understandings for cutting-edge biomedical and other miscellaneous applications as active ingredients for food formulation.

Antimicrobial Potential of Biosurfactants

The hunt for novel antimicrobial molecules is of increasing concern at the present time for the newly emerged antibiotic resistant pathogens to existing conventional antibiotics (Hancock et al. 2000). Metabolites of microbial origin have been acknowledged as a chief source of molecules capable of ingenious structural attributes with significant biological activities (Donadio et al. 2002; Banat et al. 2000; Singh and Cameotra 2004).

Antiviral potential of biosurfactants has been reported. The significant inactivation of viruses puts forward that inhibitory action is probably due to interactions between the envelope of the virus and the biosurfactant (Vollenbroich et al. 1997).

Some reports have shown significant anti-mycoplasma activities by various biosurfactants like surfactins. Contamination due to the mycoplasma in cell culture laboratories is a commonly arising serious problem when it affects the cell lines (Vollenbroich et al. 1997).

The antifungal potential of biosurfactants has long been explored (Abalos et al. 2001; Chung et al. 2000). Present microbial inhibition approaches are established on planktonic bacterial cell behavior physiology which have inadequate efficacy in terms of growth inhibition of biofilm populations. Biosurfactants have interest as anti-biofilm agents due to their potential to disperse microbial communities. In addition to the above discussed applications of biosurfactants there are several other sectors to explore biosurfactants, such as food and feed processing, active food ingredients, and green surface cleaning agents for food and biomedical surfaces

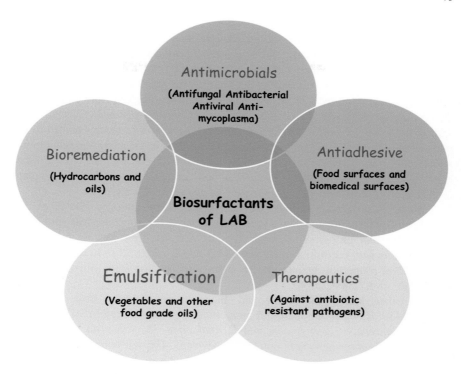

Fig. 6.1 Various applications of LAB derived biosurfactants

(Fig. 6.1). In the subsequent sections of this chapter, the potential of biosurfactants obtained from the microbial origin, particularly biosurfactants derived from probiotic organisms, will be discussed.

Antibacterial Activity

The antimicrobial property of biosurfactants is owing to the potential of molecules to self-associate and create a pore network inside cellular membrane (Deleu et al. 2008). Biosurfactants can breach into the plasma membrane through hydrophobic edges, thus ultimately manipulating the organization of the hydrocarbon chains and also fluctuating the plasma membrane thickness (Bonmatin et al. 2003). Biosurfactant mediated membrane distractions are broad-spectrum in mode of action and are beneficial for disruption of plasma membranes of pathogenic bacteria (Lu et al. 2007). Biosurfactants derived from various microorganisms showed potential antimicrobial properties (Sambanthamoorthy et al. 2014; Gudiña et al. 2015).

Rhamnolipids isolated from the *P. aeruginosa* LBI have shown significant antimicrobial properties against various bacterial and fungal pathogens, comprising *B. cereus*, *S. aureus*, *M. luteus*, and *Neurospora crassa* (Nitschke et al. 2010).

The antimicrobial property of the biosurfactant obtained from *L. paracasei* ssp. *paracasei* A20 was evaluated by determining the growth inhibition for various microorganisms (Gudiña et al. 2010). The biosurfactant derived from *L. paracasei* ssp. *paracasei* A20 was found effective against various pathogens tested. With respect to the *Lactobacillus* strains and various *Streptococcus* species related to the oral cavity such as *S. sanguis*, *S. mutans*, and *S. oralis*), broad growth reduction was witnessed for biosurfactant concentration ranges of 25–50 mg/mL. On the other hand, *E. coli*, *P. aeruginosa*, *S. aureus*, and *S. agalactiae*) showed complete reduction with biosurfactant concentration ranges of 25–50 mg/mL.

Various biosurfactants derived from LAB exhibiting antimicrobial properties have been formerly documented (Table 6.1). Conversely, there are less studies on the antimicrobial properties of biosurfactants derived from various LAB; only biosurfactants derived from *S. thermophilus* A and *L. lactis* 53 showed noteworthy antimicrobial properties against various pathogens isolated from voice prostheses (Rodrigues et al. 2004, 2006).

In another study, the cell-bound biosurfactant derived from *L. agilis* CCUG31450 showed antimicrobial properties at a concentration of 5 mg/mL against *S. aureus*, *P. aeruginosa*, and *S. agalactiae*. However, no antimicrobial activity was observed against *E. coli* and *C. albicans*. The cell-bound biosurfactants isolated from probiotic lactobacilli bacteria exhibited antibacterial properties and inhibited the growth of various drug-resistant pathogens. The biosurfactants derived from *L. jensenii* and *L. rhamnosus* were evaluated against clinical isolates of MDR *A. baumannii*, *E. coli*, and *S. aureus* and it was found that biosurfactants obtained from *L. jensenii* and *L. rhamnosus* were effective in inhibiting MDR pathogens at the concentration of 50 mg/mL. *L. jensenii* derived biosurfactant exhibited nearly 100 % inhibition against all the pathogens screened. The activity of *L. rhamnosus* oscillated from 96 to 97 % against *A. baumannii* and approx. 72–85 % against *E. coli* (Sambanthamoorthy et al. 2014).

Inhibition of microbial growth was reported against *S. enterica* and *L. mono-cytogenes* when cell free biosurfactant from *L. plantarum* and *L. Pentosus* were evaluated, displaying the maximal antimicrobial activity of cell free biosurfactants from fermentation (Rodríguez-Pazo 2013).

The initiation of antibiotics for inhibiting bacterial pathogens is regarded as one of the vital developments in contemporary medicine. The use and exploitation of antibiotics in the past 10–20 years show that the majority of clinical pathogens have established resistance to multiple antibiotics. As a result, it is necessary to find out novel antimicrobials that are potent to control infectious diseases instigated by multi drug-resistant pathogen microorganisms (Adegoke et al. 2011). Biosurfactants are increasingly seen as an incipient class of novel antimicrobial compounds. They are an appropriate alternative to conventional antibiotics and could be used as safe and potent therapeutic compounds, particularly multi-drug resistant among clinical pathogens (Singh et al. 2014).

Table 6.1 Different strains of LAB reported for antibacterial properties

S. no	Strain	Activity	References
1.	*Lactococcus lactis* 53	Exhibited significant antimicrobial activity against *Staphylococcus epidermidis, Streptococcus salivarius, S. aureus, Candida tropicalis* and *C. albicans*	Rodrigues et al. (2004, 2006)
2.	*L. plantarum* CFR2194	*E. coli, S. aureus* and *Yersinia enterocolitica*	Madhu and Prapulla (2014)
3.	*Streptococcus thermophiles* A	Antimicrobial activity against the *Candida tropicalis*	Rodrigues et al. (2006)
4.	*Lactobacillus casei*	Antimicrobial activity against *Staphylococcus aureus, Bacillus subtilis* and *Micrococcus roseus*	Golek et al. (2009)
5.	*Lactococcus lactis*	Antimicrobial activity of biosurfactant (Xylolipids) against the multi-drug resistant *Staphylococcus aureus* and *E. coli*	Saravanakumari and Mani (2010)
6.	*Lactobacillus paracasei*	Growth inhibition of *E. coli, S. agalactiae* and *S. pyogenes* with a concentration of 25 mg/ml	Gudiña et al. (2010)
7.	*Lactobacillus paracasei* A20	Antimicrobial activity against various gram positive and gram negative microorganisms at various concentration ranging from 3.12 mg/ml to 50 mg/ml	Gudiña et al. (2010)
8.	*Lactobacillus casei* MRTL3	Antimicrobial activity against *Staphylococcus aureus* ATCC 6538P, *S. epidermidis* ATCC 12228, *Bacillus cereus* ATCC 11770, *Listeria monocytogenes* MTCC 657, and *L. innocua* ATCC 33090, *Shigella flexneri* ATCC 9199, *Salmonella typhi* MTCC 733	Sharma and Saharan (2014)
9.	*L. jensenii* and *L. rhamnosus*	MDR *A. baumannii, E. coli* and *S. aureus*	Sambanthamoorthy et al. (2014)

Control of evolving multi-drug resistant pathogens is one of the main concerns in controlling opportunistic pathogens. A novel glycolipid, i.e., xylolipid biosurfactant derived from *L. lactis* and evaluation of its antibacterial property against multi-drug resistant pathogens has been studied (Saravanakumari and Mani 2010). Purified biosurfactants from *L. lactis* displayed antimicrobial properties against *E. coli* (Cd^R, Nx^R, G^R, C^R, Cz^R, K^R, Na^R, A^R, P^R, Ak^S, E^S), and *S. aureus* (Cd^S, K^R, Cz^R, P^R, G^R, A^R, C^S, Nx^R Na^R, M^R, Ak^S, E^S, Te^S, Ti^S). In well-diffusion method, biosurfactant derived from *L. lactis* considerably inhibited the growth of Methicillin resistant *S. aureus* with 12.6 mm and 13.8 mm of diameter against *E. coli*. So the biosurfactant derived from *L. lactis* can be acclaimed as a broad-spectrum antimicrobial. Biosurfactant of *L. lactis* is safe for therapeutic agent/food additive/in cosmetics, due to its probiotic origin. Antimicrobial potential of biosurfactants derived from LAB may open up the potential for biosurfactants-based alternative therapeutics for the prevention of nosocomial infections.

Anti-adhesion Activity

Microorganisms commonly settle toward solid surfaces developing biofilms as an approach to protect the planktonic cells from environmental defies (Pereira et al. 2007; Liu et al. 2012). Present microbial control approaches based on planktonic cell have been recognized to have inadequate effectiveness on biofilm communities. This is the aggravated problem due to emerging clinically important pathogens. Surfactant molecules of microbial origin have earned increasing attention in the health sector due to their potential to control microbial biofilms. The dispersal properties of biosurfactants makes them appropriate contenders for use in novel biofilm dispersal substances.

Microbial biofilms that are composed of one kind of population are comparatively sporadic in the natural environment; fairly, microorganisms are likely to be established in composite multispecies communities (Stoodley et al. 2002; Bueno 2014; Kotulova and Slobodnikova 2010). By contrast, the biofilm composition has multicellular differentiation, multicellular communication, inner architecture, and elementary fluid transport systems (Wright et al. 2005). Microbial biofilms have inconstant levels of nutrients, gas exchange, and consequently sluggish growth. These conditions in bacterial physiology can be perilous exclusively in clinical circumstances where there is a higher production of virulence factors in clinical pathogens (Sepandj et al. 2004).

The most encouraging strategy for the control of microbial biofilms have derived from biosurfactants (Sharma et al. 2015; Kiran et al. 2010). Biosurfactants derived from lactic acid bacteria have been reported with significant anti-adhesive and biofilm disruption characteristics (Sharma et al. 2014, 2015; Rodrigues et al. 2006; Gudiña et al. 2010; Dusane et al. 2010). They have been acknowledged for various potential applications in industries including agriculture, food processing, cosmetic formulations, and pharmaceutical developments (Banat et al. 2010). Microbial biofilms development on medical devices is a vital and hazardous incidence, especially pathogens resistant to existing antibiotics. Various tactics have been developed in this direction to prevent pathogen colonization. Strict hygienic practices such as routine disinfection of medical devices become important (Falagas and Makris 2009).

Approaches for the control of microbial biofilm establishment on silicone rubber voice prostheses or urethral catheters have also been documented (Rodrigues et al. 2004). Along with treatment of medical devices, microbial surfactants have been used in pretreatment of food-processing surfaces (Kim et al. 2006). The pre-conditioning of surfaces by biosurfactants could be a motivating approach to control adhesion of food associated pathogens to surfaces.

The pre-conditioning of stainless steel food surface and poly tetra-fluoro-ethylene surfaces with biosurfactant derived from L. helveticus reduced the population of L. monocytogenes (Meylheuc et al. 2006). Probiotic bacteria have long been recognized for their potential antimicrobial activity to interrupt the adhesion and expansion of biofilms to epithelial cells of urogenital and intestinal tracts (Reid et al. 2001) and the mechanisms of this interruption have been known for the release of biosurfactants from probiotic lactic acid bacteria (Gudiña et al. 2010; Reid 2002; Gupta and Garg 2009).

Biosurfactants derived from different *L. acidophilus* strains inhibited *Staphylococcus epidermidis* and *S. aureus* biofilm formation (Walencka et al. 2008). An additional exciting application for probiotics that is attaining more attention is their application in preventing oral infections (Meurman and Stamatova 2007). It has been reported that biosurfactant obtained from *Streptococcus mitis* inhibited adhesion of *Streptococcus sobrinus* and *Streptococcus mutans* to naked enamel and salivary pellicles (Van Hoogmoed et al. 2004).

In conclusion, the anti-adhesive properties of microbial surfactants against various pathogens specifies their possible effectiveness as coating agents for biomedical devices in preventing hospital acquired infections without the prerequisite for use of conventional drugs and antimicrobials.

Conclusion

Metabolites of microbial origin have been acknowledged as a chief source of molecules possessing ingenious structural attributes with significant biological activities. Biosurfactants have interest as antibiofilm agents due to their potential to disperse microbial communities. Biosurfactant mediated membrane distractions are broad-spectrum in mode of action and are beneficial for disruption of plasma membranes of pathogenic bacteria. Control of evolving multi-drug resistant pathogens is one of the main necessities in controlling opportunistic pathogens. Antimicrobial potential of biosurfactants derived from LAB may open up the potential for biosurfactants-based alternative therapeutic for the prevention of nosocomial infections. Biosurfactants derived from lactic acid bacteria have been reported with significant anti-adhesive and biofilm disruption characteristics.

References

Abalos A, Pinazo A, Infante MR, Casals M, Garcia F, Manresa A (2001) Physicochemical and antimicrobial properties of new rhamnolipids produced by *Pseudomonas aeruginosa* AT10 from soybean oil refinery wastes. Langmuir 17(5):1367–1371

Abazov VM, Abbott B, Acharya BS, Adams M, Adams T, Alexeev GD, Borissov G (2011) Forward-backward asymmetry in top quark-antiquark production. Phys Rev D 84(11):112005

Adegoke AA, Iberi PA, Akinpelu DA, Aiyegoro OA, Mboto CI (2011) Studies on phytochemical screening and antimicrobial potentials of Phyllanthus amarus against multiple antibiotic resistant bacteria. Int J Appl Res Nat Prod 3(3):6–12

Banat IM, Makkar RS, Cameotra SS (2000) Potential commercial applications of microbial surfactants. Appl Microbiol Biotechnol 53(5):495–508

Banat IM, Franzetti A, Gandolfi I, Bestetti G, Martinotti MG, Fracchia L, Marchant R (2010) Microbial biosurfactants production, applications and future potential. Appl Microbiol Biotechnol 87(2):427–444

Bharali P, Saikia JP, Ray A, Konwar BK (2013) Rhamnolipid (RL) from *Pseudomonas aeruginosa* OBP1: a novel chemotaxis and antibacterial agent. Colloids Surf B: Biointerfaces 103:502–509

Bonmatin JM, Laprévote O, Peypoux F (2003) Diversity among microbial cyclic lipopeptides: iturins and surfactins. Activity-structure relationships to design new bioactive agents. Comb Chem High Throughput Screen 6(6):541–556

Bueno CS (2014) Curso sistematizado de direito processual civil

Chung YR, Kim CH, Hwang I, Chun J (2000) Paenibacillus koreensis sp. nov., a new species that produces an iturin-like antifungal compound. Int J Syst Evol Microbiol 50(4):1495–1500

Deleu M, Paquot M, Nylander T (2008) Effect of fengycin, a lipopeptide produced by *Bacillus subtilis*, on model biomembranes. Biophys J 94(7):2667–2679

Donadio S, Monciardini P, Alduina R, Mazza P, Chiocchini C, Cavaletti L, Puglia AM (2002) Microbial technologies for the discovery of novel bioactive metabolites. J Biotechnol 99 (3):187–198

Dusane DH, Zinjarde SS, Venugopalan VP, Mclean RJ, Weber MM, Rahman PK (2010) Quorum sensing: implications on rhamnolipid biosurfactant production. Biotechnol Gen Eng Rev 27 (1):159–184

Falagas ME, Makris GC (2009) Probiotic bacteria and biosurfactants for nosocomial infection control: a hypothesis. J Hosp Infect 71(4):301–306

Golek P, Bednarski W, Brzozowski B, Dziuba B (2009) The obtaining and properties of biosurfactants synthesized by bacteria of the genus *Lactobacillus*. Ann Microbiol 59 (1):119–126

Gudiña EJ, Rocha V, Teixeira JA, Rodrigues LR (2010) Antimicrobial and antiadhesive properties of a biosurfactant isolated from *Lactobacillus paracasei* ssp. *paracasei* A20. Lett Appl Microbiol 50(4):419–424

Gudiña EJ, Fernandes EC, Rodrigues AI, Teixeira JA, Rodrigues LR (2015) Biosurfactant production by *Bacillus subtilis* using corn steep liquor as culture medium. Frontiers Microbiol 6

Gupta V, Garg R (2009) Probiotics. Indian J Med Microbiol 27(3):202

Hancock D, Gay J, Rice D, Davis M, Gay C, Besser T (2000) The global epidemiology of multiresistant Salmonella enterica serovar Typhimurium DT104

Kim HS, Jeon JW, Kim BH, Ahn CY, Oh HM, Yoon BD (2006) Extracellular production of a glycolipid biosurfactant, mannosylerythritol lipid, by *Candida* sp. SY16 using fed-batch fermentation. Appl Microbiol Biotechnol 70(4):391–396

Kiran GS, Thomas TA, Selvin J, Sabarathnam B, Lipton AP (2010). Optimization and characterization of a new lipopeptide biosurfactant produced by marine *Brevibacterium aureum* MSA13 in solid state culture. Bioresour Technol 101(7):2389–2396

Kotulova D, Slobodnikova L (2010) Susceptibility of Staphylococcus aureus biofilms to vancomycin, gemtamicin and rifampin. Epidemiologie, mikrobiologie, imunologie: casopis Spolecnosti pro epidemiologii a mikrobiologii Ceske lekarske spolecnosti JE Purkyne 59 (2):80–87

Liu B et al (2012) Deep sequencing of the oral microbiome reveals signatures of periodontal disease. PloS one 7(6), e37919

Lu AH, Salabas EE, Schüth F (2007) Magnetic nanoparticles: synthesis, protection, functionalization, and application. Angew Chem Int Ed 46(8):1222–1244

Madhu AN, Prapulla SG (2014) Evaluation and functional characterization of a biosurfactant produced by Lactobacillus plantarum CFR 2194. Appl Biochem Biotechnol 172(4):1777–1789

Marchant R, Banat IM (2012) Biosurfactants: a sustainable replacement for chemical surfactants? Biotechnol Lett 34(9):1597–1605

Meurman JH, Stamatova I (2007) Probiotics: contributions to oral health. Oral Dis 13(5):443–451

Meylheuc T, Methivier C, Renault M, Herry JM, Pradier CM, Bellon-Fontaine MN (2006) Adsorption on stainless steel surfaces of biosurfactants produced by gram-negative and gram-positive bacteria: consequence on the bioadhesive behavior of *Listeria monocytogenes*. Colloids Surf B 52(2):128–137

Neu TR (1996) Significance of bacterial surface-active compounds in interaction of bacteria with interfaces. Microbiol Rev 60(1):151

Nitschke M, Costa SG, Contiero J (2010) Structure and applications of a rhamnolipid surfactant produced in soybean oil waste. Appl Biochem Biotechnol 160(7):2066–2074

Pereira VJ, Linden KG, Weinberg HS (2007) Evaluation of UV irradiation for photolytic and oxidative degradation of pharmaceutical compounds in water. Water Res 41(19):4413–4423

Reid G, Zalai C, Gardiner G (2001) Urogenital lactobacilli probiotics, reliability, and regulatory issues. J Dairy Sci 84:E164–E169

Reid G (2002) The potential role of probiotics in pediatric urology. J Urol 168(4):1512–1517

Rodrigues L, Van der Mei H, Teixeira JA, Oliveira R (2004) Biosurfactant from *Lactococcus lactis* 53 inhibits microbial adhesion on silicone rubber. Appl Microbiol Biotechnol 66 (3):306–311

Rodrigues L, Banat IM, Teixeira J, Oliveira R (2006) Biosurfactants: potential applications in medicine. J Antimicrobial Chemother, 57(4):609–618

Rodríguez-Pazo N, Salgado JM, Cortés-Diéguez S, Domínguez JM (2013) Biotechnological production of phenyllactic acid and biosurfactants from trimming vine shoot hydrolyzates by microbial coculture fermentation. Appl Biochem Biotechnol 169(7):2175–2188

Rosenberg E, Ron EZ (1997) Bioemulsans: microbial polymeric emulsifiers. Curr Opin Biotechnol 8(3):313–316

Sambanthamoorthy K, Feng X, Patel R, Patel S, Paranavitana C (2014) Antimicrobial and antibiofilm potential of biosurfactants isolated from lactobacilli against multi-drug-resistant pathogens. BMC microbiol 14(1):1

Saravanakumari P, Mani K (2010) Structural characterization of a novel xylolipid biosurfactant from *Lactococcus lactis* and analysis of antibacterial activity against multi-drug resistant pathogens. Bioresour Technol 101(22):8851–8854

Sepandj F, Ceri H, Gibb A, Read R, Olson M (2004) Minimum inhibitory concentration (MIC) versus minimum biofilm eliminating concentration (MBEC) in evaluation of antibiotic sensitivity of gram-negative bacilli causing peritonitis. Perit Dial Int 24(1):65–67

Sharma D, Singh Saharan B (2014) Simultaneous Production of biosurfactants and bacteriocins by probiotic *Lactobacillus casei* MRTL3. Int J Microbiol 2014

Sharma D, Saharan BS, Chauhan N, Procha S, Lal, S (2015) Isolation and functional characterization of novel biosurfactant produced by *Enterococcus faecium*. SpringerPlus 4 (1):1–14

Singh P, Cameotra SS (2004) Potential applications of microbial surfactants in biomedical sciences. Trends Biotechnol 22(3):142–146

Singh AK, Rautela R, Cameotra SS (2014) Substrate dependent in vitro antifungal activity of *Bacillus* sp strain AR2. Microb Cell Fact 13(1):1

Smyth TJP, Perfumo A, McClean S, Marchant R, Banat IM (2010) Isolation and analysis of lipopeptides and high molecular weight biosurfactants. In: Handbook of hydrocarbon and lipid microbiology. Springer, Berlin Heidelberg, pp 3687–3704

Stoodley P, Sauer K, Davies DG, Costerton JW (2002) Biofilms as complex differentiated communities. Annu Rev Microbiol 56(1):187–209

Van Hamme JD, Singh A, Ward OP (2006) Physiological aspects: Part 1 in a series of papers devoted to surfactants in microbiology and biotechnology. Biotechnol Adv 24(6):604–620

Van Hoogmoed LM, Nieto JE, Snyder JR, Harmon FA (2004) Survey of prokinetic use in horses with gastrointestinal injury. Vet Surg 33(3):279–285

Vollenbroich D, Özel M, Vater J, Kamp RM, Pauli G (1997) Mechanism of inactivation of enveloped viruses by the biosurfactant surfactin from *Bacillus subtilis*. Biologicals 25 (3):289–297

Walencka E, Różalska S, Sadowska B, Różalska B (2008) The influence of Lactobacillus acidophilus-derived surfactants on staphylococcal adhesion and biofilm formation. Folia Microbiol 53(1):61–66

Wright KJ, Higgs DM, Belanger AJ, Leis JM (2005) Auditory and olfactory abilities of pre-settlement larvae and post-settlement juveniles of a coral reef damselfish (Pisces: Pomacentridae). Mar Biol 147(6):1425–1434

Chapter 7
Future Prospect

Abstract Increasing interest in green and sustainable solutions in industrial and food formulation derives the force behind the rising number of studies from probiotic microorganisms. Biosurfactants obtained from probiotic lactic acid bacteria could be used as future cosmetics base, food additives, and therapeutics molecules. Further advancement in structural and metabolic engineering accumulates information about the structural properties of biosurfactants and this information could be useful to develop tailor-made biosurfactants for future applications such as green capping of nanoparticles. The future of biosurfactant application to human health would be an area of interest for various research groups around the globe.

Keywords Cosmetics · Nanoparticles · Food · Feed · Biosurfactants

Introduction

As witnessed by the increasing research and the rising number of publications on biosurfactants and their applications, there is increasing concern over the development of these amphiphilic molecules. The plea for novel biosurfants in cosmetic formulations, food processing, pharmaceutical base, and environmental applications is progressively growing. The recognized biomedical properties of biosurfactants and the recent applications in food, synthesis of nanomaterial, and cosmetics formulations recommend that it is worth continuing the biosurfactant research. However, the production cost is a limitation of commercialization of biosurfactants. Besides, in food, pharmaceutical preparation, and biomedical applications, the expenses incurred in production, i.e., high production cost could be reduced for by the small quantities of biosurfactants consumed. Requirements are ongoing for making biosurfactant production more cost-effective with feasible production conditions and competent product recovery, as well as the use of certain hyperproducing strains.

Current developments in the field of biomedical application of biosurfactants are undoubtedly increasing due to higher prospective economic revenues. Furthermore, with outstanding self-assembly activity, innovative and interesting solicitations in nanotechnology are anticipated for biosurfactants research. In-depth research of biosurfactant natural roles in microbial interactions, cell-to-cell signaling, biofilm formation, pharmaceutical formulations, and food processing could propose better and motivating future applications.

Biosurfactant Mediated Synthesis of Nanoparticles

The prerequisites of novel greener bioprocessing advances for synthesis of nanoparticles are evolving as a replacement strategy. The biosurfactant arbitrated practice and microbial synthesis of metal nanoparticles are currently developing as clean, nontoxic, and environmentally safe "green preparation" techniques. Surface active agent mediated nanoparticle synthesis is superior to approaches of using bacterial or fungal-based nanoparticle synthesis, as biosurfactants reduce the establishment of aggregates and expedite a uniform morphology for synthesized nanoparticles.

Microbial surfactant micelles arise in different morphologies, such as spherical, ellipsoidal, and cylindrical structures. The main dread in nanoparticle synthesis is the change in these assemblies to attain aggregates with desired structure. The experimental setup such as change in pH, temperature, capping agents with metal ions, and viscosity of solution are a few facets that can be particular in the production and design of nanoparticles.

Predominantly, capping agents play a precisely significant role in defining the final quality of metal nanoparticles. It basically decreases the affinity of metal nanoparticles to agglomerate, by shielding the surface by initiating electrostatic stabilization. Biosurfactants, as capping agents in metal nanoparticle production, also enable the constant dispersion of metal nanoparticles in the liquid medium. Biosurfactants as dispersion agents showed to be exactly effective for nanoparticles by rhamnolipids and sophorolipids. It can reduce the problem of environmental vulnerabilities and that of expensive organic solvents and fatty acids used as capping agents.

Mutually, biosurfactants and metal nanoparticles have an extensive variety of medical, food, environmental, and agricultural applications. The biosurfactant-facilitated metal nanoparticle synthesis is an effective, environmentally, and safe benign procedure, with further research prerequisites to be engrossed on the collective effect of biosurfactant and metal nanoparticles which would finally bring out new progressions.

Biosurfactants in Animal Feed and Food

Although having noteworthy potential concomitant with emulsion formation, stabilization, antibiofilm, and antimicrobial properties, considerably less applications have been documented in the food processing industry. Emulsification plays a vital role in consistency, texture, and solubilization of aromas in food formulations. The purpose of an emulsifier is to stabilize emulsion by adjusting the clustering of fat globules. Stability can be improved by addition of surfactants, which decrease the interfacial tension through the establishment of electrostatic barriers.

In bakery and ice cream formulations, microbial surfactants have been documented to develop consistency, adjourning staling, and solubilizing oils and as anti-spattering agents and fat stabilizers. Rhamnolipids glycolipid, specifically, have been applied to improve dough texture and stability. L-Rhamnose is significant as a precursor for flavor components like furaneol. Increasingly, food industries have been steadily looking to lessen dependency on plant-based emulsifiers derived from genetically modified crops. So it is hereby concluded that through focused inexpensive application of biosurfactants in food processing and focused testing and examinations of toxicity, we can look presumptuously at microbial surfactants as the additives of the future.

Biosurfactants in Cosmetics

Surfactants in cosmetic formulations accomplish detergency, emulsifying, dispersing, and foaming properties. Contrary reactions of chemical originated surfactants have an effect on human health, predominantly severe in long-term exposure. Microbial surfactants derived from different microorganisms demonstrating potential surface properties fit for cosmetic formulations specifically integrate with their biological activities. Several glycolipids biosurfactants such as sophorolipids, rhamnolipids, and mannosylerythritol lipids are the most extensively applied biosurfactants in cosmetics. Research and patents appropriate to various glycolipids studied highlight the cosmetic applications comprising personal healthcare formulations offering cosmetic effectiveness, value, and economy benefits of biosurfactants.

Conclusion

Biosurfactants are the molecules of choice for the future prospective in food formulations, nanoparticles synthesis, biomedical, and healthcare services. Further research to minimize the production and operational cost of production of biosurfactants will strengthen the demand for surface active agents. Toxicity evaluation and better compatibility assessment of the biosurfactant with human and animal-related material are the need of the hour to develop biosurfactant-mediated processes. Further study and better structural evaluation of the biosurfactant are required for industrial development and for future applications.